Intelligent Systems, Control and Automation: Science and Engineering

Volume 89

More information about this series at http://www.springer.com/series/6259

Harutoshi Ogai · Bishakh Bhattacharya

Pipe Inspection Robots for Structural Health and Condition Monitoring

Springer

Harutoshi Ogai
Graduate School of Information,
Production and Systems (IPS)
Waseda University
Kitakyushu, Fukuoka
Japan

Bishakh Bhattacharya
Department of Mechanical Engineering
Indian Institute of Technology Kanpur
Kanpur, Uttar Pradesh
India

ISSN 2213-8986 ISSN 2213-8994 (electronic)
Intelligent Systems, Control and Automation: Science and Engineering
ISBN 978-81-322-3749-5 ISBN 978-81-322-3751-8 (eBook)
https://doi.org/10.1007/978-81-322-3751-8

Library of Congress Control Number: 2017954471

Printed on acid-free paper

This Springer imprint is published by Springer Nature
The registered company is Springer (India) Private Ltd.
The registered company address is: 7th Floor, Vijaya Building, 17 Barakhamba Road, New Delhi 110 001, India

Dedicated
to
The Students of Japan and India
engaged in the field of Robotics
and Automation

Preface

The chance to write this textbook happened during the joint research and technical exchanges on pipe inspection robots by the two authors, Prof. Bishakh Bhattacharya from Indian Institute of Technology Kanpur (IITK) and Prof. Harutoshi Ogai from Graduate School of Information, Production and Systems (IPS), Waseda University.

Ogai came across upon a robot with integrated smart sensor for the piping inspection in the laboratory of Dr. Bhattacharya when he visited the Department of Mechanical Engineering, IITK, in 2008. At that time, while Ogai was focusing on the research of wireless sensor networks for bridge health monitoring system, a local Kitakyushu research team was developing an inspection robot for sewage pipe inspection (Kantarou). It was considered a research challenge on how to make the robot's radio communication system for movement inside the pipe. As described in this book, the radio characteristics inside the pipe were measured, and a type of wireless communication system was developed based on the measurement results. Also, the health monitoring system for the pipe was developed by using a video camera.

During this time, IITK had developed a pipe inspection robot that performed mechanical defect detection by touching pipe surfaces. The choice of touch-based sensor development was deliberate to avoid high-density data generation which could have created storage problem inside the robot.

Under such milieu, the two laboratories started extensive collaborative researches and student exchanges and carried out joint research on pipe inspection robots. They jointly developed a new prototype robot using both camera and mechanical touch sensing for inspection. Currently, the two laboratories are working together for developing pipe inspection robots for various applications including oil and gas pipelines.

The design, development, and experiments on the wireless communication systems discussed in this book were carried out jointly with the help of Hakutsu Technology Corporation, which contributed to the research and development of such systems in Kitakyushu now. As for the production of the main body of the robot, our sincere thanks are due to the cooperation of the Ishikawa Iron Works.

By showing some of the research achievements, this book is aimed at providing some reference and sharing experience on designing and producing pipe inspection systems to the university students and practicing engineers. As a reference to the actual pipe inspection, the book discusses the present condition of the inspection robot of the sewage pipe, the present inspection state of the gas pipe inspection robot, the design of the pipe inspection robot, the wireless communication characteristics of the pipe, the wireless communication system and the camera-based pipe inspection robot, the role of smart materials, smart sensors, structural health monitoring, and future pipe inspection robot technology.

It will be our honor if the book is found to be helpful in the development of future pipe inspection technology.

Finally, great thanks to the contributions from Waseda University, Indian Institute of Technology Kanpur, Hakutsu Technology Corporation, Ishikawa Iron Works, Gas Authority of India Limited, and the students. Particularly, for wireless communications, thanks to Mr. Katsumi Hirai, Mr. Takahiko Abe, Mr. Gunkichi Sato from Hakutsu Technology Corporation, Mr. Hsiang-Ping Yang who graduated from Waseda University, Dr. Shigeyuki Tateno from Waseda University, Mr. Kaito Yano from Waseda University, Mr. Nitin Pal and Mr. Anubhaw-Kumar Jain from IITK. For the prototype robot, thanks to Mr. Kiyomitsu Ishikawa; for camera image processing, thanks to Mr. Zhicheng Wang from Ogai laboratory, Waseda University; for multi-dimensional traveling robot research, thanks to Mr. Hsiang-Pin Yang, Mr. Chen Chu Kang, Mr. Zhiqiang Tang who graduated from Waseda University and Mr. Zhuochao GU a student from Ogai laboratory. Also, special thanks to Prof. Hiroshi Inujima from Waseda University, Dr. Wa Si, Mr. Kentarou Nishijima, and Mr. Ryuta Oyabu who graduated from Ogai laboratory, Waseda University.

From the students of SMSS laboratory, our deep appreciation goes to Mr. Nayan Jyoti Vaishya, Mr. Gaurav Bansal, Mr. Gaurav Verma, Mr. Himashu Pandey, Mr. Hardik Soni, Mrs. Sanjana Chakraborty, and Mr. Aravind Kumar who have helped enormously to make this book informative and presentable. The authors are grateful to Prof. P.K. Panigrahi, Head, Mechanical Engineering, and Prof. K. Muralidhar, Dean of Faculty Affairs, IITK along with Dr. and Mrs. Mehta Chair, for actively supporting many visits of Prof. Ogai, which was crucial for writing the book.

In the personal front, Bishakh humbly takes the opportunity to acknowledge the constant inspiration and editing help from his wife Chitralekha, son Chandramouli, and little daughter Bipasana. Without their sacrifice, the writing would have been impossible at a difficult time. Bishakh has been eternally blessed and inspired by his father late Bholanath Bhattacharya, mother Bina Bhattacharya, father-in-law Kamalesh Banerjee, and mother-in-law Maya Banerjee for selecting academics as the sole goal for persuasion in life. This book is a little milestone toward that direction.

Kitakyushu, Japan
Kanpur, India
July 2017

Harutoshi Ogai
Bishakh Bhattacharya

Contents

About the Authors

Prof. Harutoshi Ogai holds a Doctorate in Engineering from the Tokyo Institute of Technology, Japan. He is currently a Professor at the Graduate School of Information, Production and Systems (IPS), Waseda University, Japan. Before joining the IPS, Waseda University, he had worked as one of the chief executives with Nippon Steel Company for over two decades and had made seminal contributions to the field of modeling and optimal control of complex systems like blast furnaces. His current research areas include complex process modeling and control, design and development of autonomous electric vehicles, process analysis and simulation, bridge damage diagnosis, and automotive control. He has successfully developed process modeling and Java-based process simulators for many large-scale complex processes and developed strategies for optimal control, chiefly using independent component analysis (ICA), large-scale data on-line modeling (LOM), and simulation-based on-line guidance (SOG). Professor Ogai and his team have developed many such hardware-in-the-loop systems for various industrial applications.

Dr. Bishakh Bhattacharya is the Mehta Chair Professor, Department of Mechanical Engineering, Indian Institute of Technology Kanpur (IITK), India. Among other activities, he was the Head of the Design Programme from 2011 to 2013. He is currently coordinating the Space Technology Cell of the Institute. He is active in creating Web- and video-based course materials for India's National Program on Technology Enhanced Learning (NPTEL) initiative. He was a Visiting Professor at the University of Sheffield, where he carried out research on structural health monitoring of complex autonomous systems. He also visited IPS, Waseda University, Japan, to jointly develop new pipe health monitoring robots. His key areas of research include vibration control, structural health monitoring, energy harvesting system, control–structure interaction, intelligent system design, and child–robot Interaction. Based on his work, a new technology is currently being developed for the shape control of continuous structures like space-born reconfigurable parabolic antenna systems for the Indian Space Research Organization (ISRO). Another notable application of his work is a new robot for pipe health monitoring, developed for the Gas Authority of India Limited (GAIL). Professor Bhattacharya is a recipient of the Department of Science and Technology (DST) Young Scientist Award and Young Scientist Award from the Systems Society of India. He has developed international collaborations with the UK, Sweden, and Japan.

Chapter 1
Pipe Inspection Robots for Sewerage Pipelines

1.1 A Broad Overview

A sewer is defined as a pipe (an artificial subterranean conduit or tunnel) that collects and transports waste and drainage water (e.g., surface runoffs such as storm water, rainwater, and groundwater). Sewage collection systems are an extensive and vital part of the infrastructure for any country.

In 2007, the total length of sewer pipes in Japan was over 400,000 km (about 10 times of the circumference of the earth). However, a report released by the Japan Sewage Works Association (JSWA) suggests that over 70,000 km of the sewer pipes (nearly one-seventh of the total length of sewer pipes in Japan) have been well over their service life (more than 50 years) [1]. Consequently, the aging problem of the sewer pipes is being taken into serious consideration.

A major part of sewer pipe deterioration and corrosion is attributed to aging, traffic load, and chemical reactions, due to which inflows such as groundwater and rainwater seep into the pipe systems. Regional city government reports state that this inflow amounts to approximately 30% of the total flow. In addition to these inflows into the sewer pipes, outflow from damaged systems also occurs, thereby contaminating the surrounding environment and posing a greater threat. The primary cause of such problems is the lack of in-depth knowledge about the condition of the pipelines. Under similar circumstances, many countries across the world have been spending a large amount of their budgets on emergency repairing of sewers. The need for emergency repairing of subterranean pipes can be significantly reduced if critical sections of it could be identified and repaired before any catastrophic failure occurs. On the other hand, the requirements of rebuilding the sewer pipes and the laying of optical fibers through them are also increasing. Hence, it is essential to investigate the condition of sewer pipes before any such construction/reconstruction.

H. Ogai and B. Bhattacharya, *Pipe Inspection Robots for Structural Health and Condition Monitoring*, Intelligent Systems, Control and Automation: Science and Engineering 89, https://doi.org/10.1007/978-81-322-3751-8_1

To put it briefly, the use of effective inspection system for sewer pipes is gradually gaining importance and is even considered in some places a mandatory procedure before any construction. This requires a convenient operational environment, a short investigation time, and an accurate fault identification system. Also, the key to developing this system is to find a fast, robust, and accurate fault diagnostics method for the sewer pipes.

1.1.1 Problem Statement

In Japan, from 1982, instead of manual inspection system, cable-tethered robot-inspectors integrated with onboard video camera have been used extensively. This becomes specifically important for small-bore sewer pipes that cannot be entered by humans directly. Such robots are connected to the outside of the sewer system by a cable. The cable is used for multiple purposes such as energy supply, transmission of commands from a human operator to the device, data transmission back to an operator, and measurement of the distance traveled. Also, it acts as a lifeline in case the device gets stuck inside the pipe. An operator remotely controls the movement of the robot and the video system. This video-supported visual inspection system records notable damages or abnormalities in the video stream facilitating the monitoring of the current condition of sewer pipes.

However, these sewer inspection systems have the following disadvantages:

- For the conventional inspection system, the condition of the sewer pipe is monitored through the video by human eyes. Hence, the reliability of this system depends entirely on the experience and state of the mind of the operator. And it is only human to make mistakes involving misjudgment or oversight of defects.
- Also, for such a system, the video stream of the sewer pipe is recorded during the investigation and is monitored later. As manually checking the video tends to be time-consuming, it automatically takes a lot of time for the completion of the whole investigation process.
- In addition, huge memory storage space is required to save the records of the video stream of sewer pipes. Added to that, an additional cost is incurred for the examination and maintenance of the sewer pipes.
- For the conventional inspection system, a cable is used to control the robot, thus limiting the distance covered to the length of the cable employed. To overcome this, a cable of sufficient length has to be carried to the investigation site, making the work inconvenient.

Consequently, building an effective autonomous robot capable of online identification and extraction of the defects from the video image is of immediate necessity.

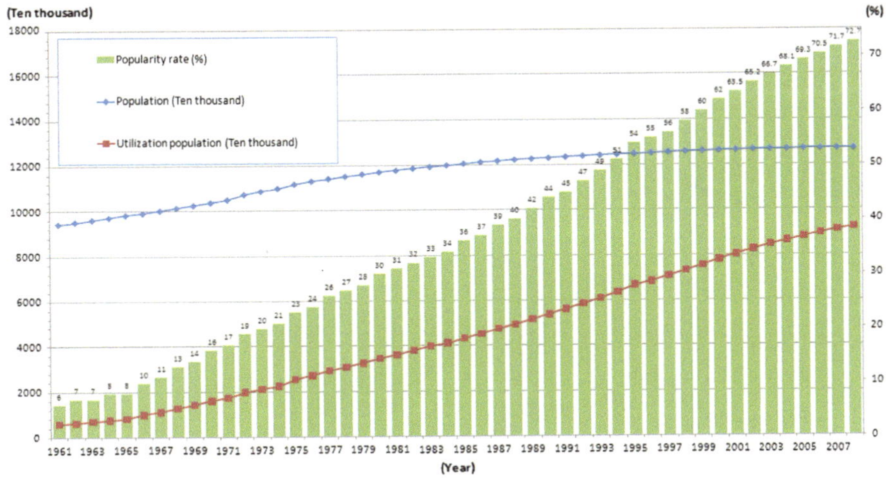

Fig. 1.1 Trend of popularity rate of the sewer pipes in Japan [2]

1.1.2 Sewer Pipes in Japan

In Japan, as sewer pipes are very important for daily life, quite a lot of new sewer pipes are being built every year. As per a report by the Japan Sewage Works Association (JSWA), the popularity rate of the sewer pipes in 2009 has been 72.7% (Fig. 1.1) [2].

The popularity rate of sewer pipes has steadily increased over the past 50 years. However, there are still some regions in Japan which have not been connected with sewer pipes, and the Japanese government has assured to resolve the crisis within the next 10 years.

On the other hand, with the increase in the total length of sewer pipes, the aging problem and other problems caused by aging of the sewer pipes are increasing rapidly. For that, it becomes increasingly necessary to gain accurate and efficient information regarding the damage in existing sewer pipes.

1.1.3 Need for Accurate Determination of Damage in Sewer Pipes

In Japan, more than 70,000 km of sewer pipes have been over their service lives (more than 50 years). Consequently, the most serious problem caused by aging of the sewerage system is the subsidence of roads (Fig. 1.2). In 2009, there were about 4700 places where such subsidence occurred and most of them were caused due to the aging of sewer pipes (Fig. 1.3).

Fig. 1.2 Subsidence of the road (Sumida District, Tokyo)

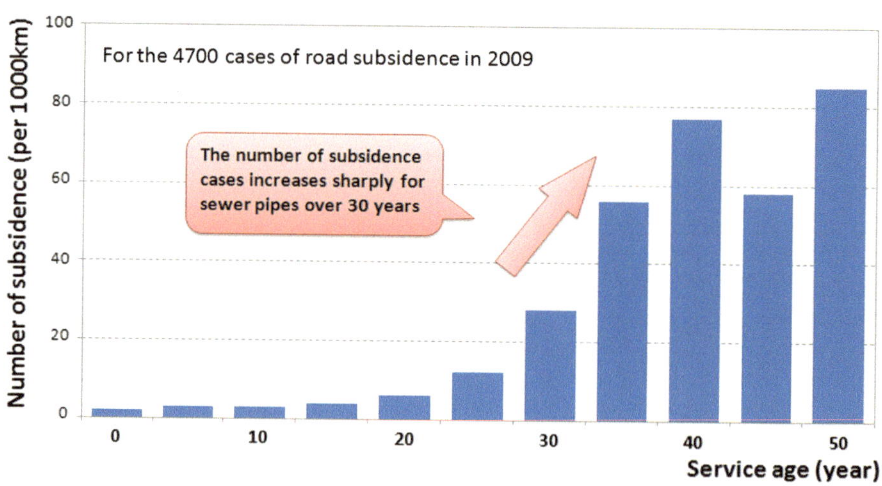

Fig. 1.3 Number of subsidence cases in 2009 [3]

On the other hand, for the energy saving and recycling society, recycled use of sewer spaces has also become a major problem. The main issue of recycling sewer space lies in the space occupied in the laying of fiber-optic cables effectively aiding information and communication network. Table 1.1 provides an overview of fiber-optic cable lines across various states of Japan.

Table 1.1 Length of fiber-optic cables for each state in Japan [4]

Development bureau	Name of the state	Length (in km)
Hokkaido	Hokkaido	89.21
Kanto	Tokyo	753.28
	Kanagawa	152.1
Chubu	Nagano	0.60
	Gifu	94.4
	Aichi	115.00
Hokuriku	Toyama	14.50
	Ishikawa	1.50
Kansai	Kyoto	18.9
	Osaka	78.2
	Hyogo	63.5
Chugoku	Okayama	196.07
	Tokushima	0.40
	Yamaguchi	0.18
Shikoku	Kochi	1.50
Kyushu	Fukuoka	9.2
	Miyazaki	0.93
	Kumamoto	0.37
	Kagoshima	1.30
Total length		1,591.14

With the aging problem and the increase in the recycling of sewer pipes, inspection of the pipes is of utmost importance. Hence, every year, an increasing number of sewer pipes needs to be investigated and the cost of these investigation works will prove to be a large burden for the government. On the other hand, due to

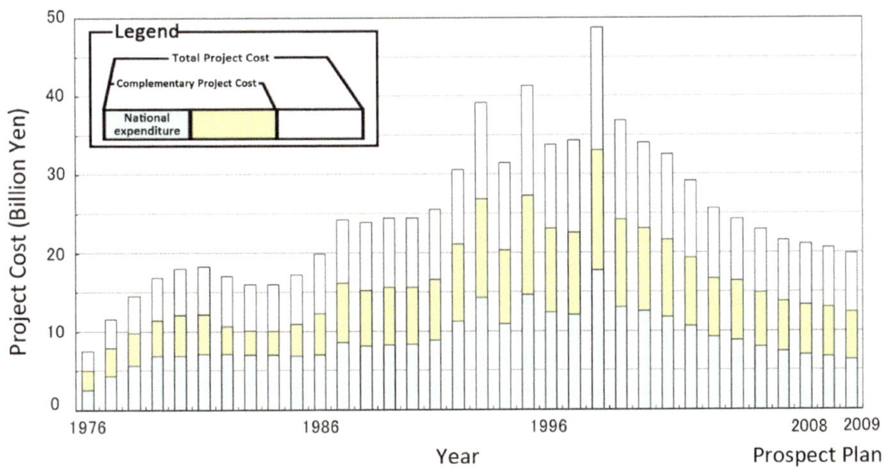

Fig. 1.4 Trend of budget for the sewerage monitoring project in Japan [3]

the economic depression, the resource allocated for the sewer pipe monitoring project has been decreasing over the years (Fig. 1.4).

Consequently, with the increasing need for the inspection of sewer pipes and the decreasing budget for the sewer pipe monitoring project, finding an efficient inspection method for sewer pipes is becoming quite challenging with time.

1.2 Functional Requirements of a Sewer Inspection Robot

From the functional point of view, Sewer Inspection Robots are of two types: passive/semiautonomous robots and fully autonomous robots. Semiautonomous robots are controlled by a human operator, and they have a limited range as they are cabled at the end. Fully autonomous robots, on the other hand, are completely capable of autonomous navigation and inspection. In order to achieve this functional requirement, they generally possess six important subsystems:

- A set of navigation sensors for motion control such as accelerometers, servomotors;
- Localization sensors for identifying the position of the robot with respect to the sewer pipe using optical/IR camera;
- An array of damage inspection sensors using camera/laser scanner;
- Actuator/wheel mechanism;
- A processing unit for sensor data storage and compilation; and
- Communication system with the base station.

A similar architecture is also used for gas-pipe inspection robots. This is described at length in the following chapter. In this chapter, we will focus specifically on the techniques of sewer inspection.

1.3 Sewer Inspection Techniques

There are various sewer inspection techniques that are used to assess the condition of sewers, and they can be broadly classified into three different groups [5]. An overview of this classification is illustrated in Fig. 1.5. The first group consists of techniques that determine the internal condition of a sewer. The commonly used examples for this group are Closed-Circuit Television (CCTV) inspections and Sewer Scanner and Evaluation Technology (SSET). The second group examines the overall condition of sewers and the surrounding soil. The third group detects specific problems within or behind a sewer wall [6]. Among all the inspection techniques, CCTV inspection is the most widely used method. Table 1.2 depicts a brief summary of the commonly used inspection techniques for assessing the existing condition of sewers. It suggests the most appropriate usage of the different techniques and highlights the key advantages and disadvantages of these techniques.

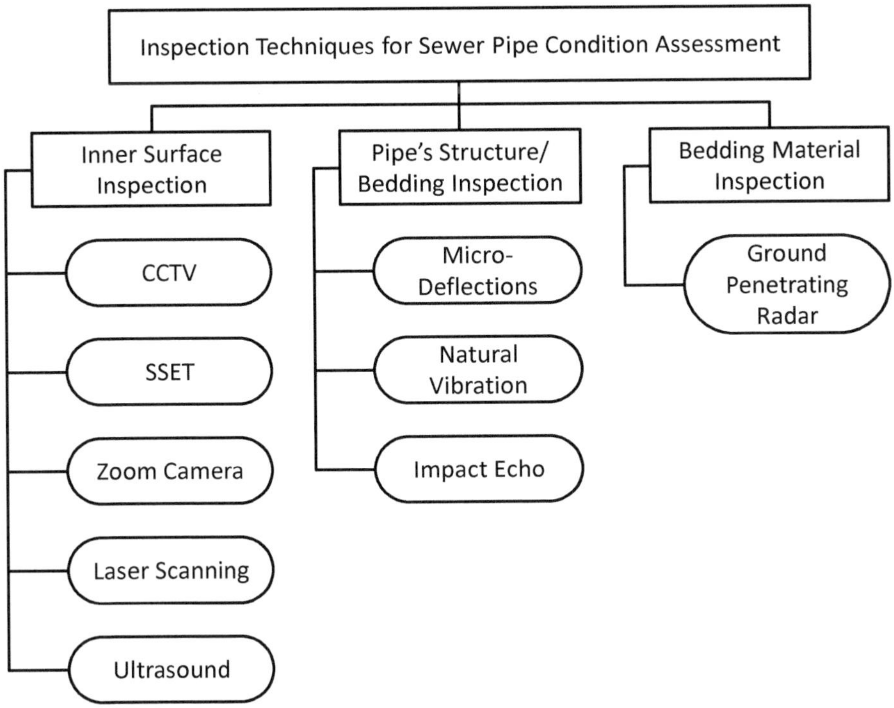

Fig. 1.5 Inspection techniques for sewer pipe condition assessment

1.3.1 CCTV Inspection Systems

Standard pipe inspection systems based on Closed-Circuit Television (CCTV) cameras have a large range of application fields including waste pipes and drains. The CCTV method consists of a mobile, remotely operated platform (usually equipped with a high-resolution color video camera and a lighting system). The camera platform is connected via a multi-core cable to a remote inspection station with video recording facilities situated on the ground. An engineer assesses the recorded images offline.

There are two basic types of CCTV system stationary CCTV system and mobile CCTV system [7]. Each uses a television camera in combination with a video monitor, video cassette recorders, and possibly other recording devices.

- **Stationary CCTV Systems**

Stationary video cameras mounted at a manhole are limited with respect to what they can see. Defects located close to the manhole can be detected, but the farther a defect lies into the sewer pipe, the harder it becomes to identify and evaluate it. Defects beyond the range of the camera would be missed entirely unless they cause secondary effects that can be identified in the manhole (such as changes in water

Table 1.2 A comparison of commonly used sewer inspection techniques [5, 6]

Inspection group	Technique	Usage	Common detections	Advantages	Disadvantages
Inner surface	CCTV	Empty and partially filled pipes	Surface cracks, deformations, erosion, infiltration, blockage	Standard technique, relatively cheap, easily available, evaluates entire length	May miss hidden defects causing different interpretations of result
	SSET	Pipes of diameter from 8 to 24 in.	As CCTV but with higher accuracy	Higher accuracy, good judgment of pipe deformation	Expensive
	Zoom camera	Empty and partially filled pipes	As CCTV but with lower accuracy	No cleaning/flushing is required before inspection, cheaper than CCTV	Cannot cover the whole length of a pipe, results are less accurate than CCTV
	Laser scanning	Empty and partially filled pipes	Surface cracks, erosion, missing bricks, deformations	Computer-based analysis, accurate defect and geometry measurement	Only works above water line
	Ultrasound	Empty to flooded pipes	Deformation, erosion, brick damage	Measures defect above and below water line, computer-based analysis	Extensive cleaning of sewer necessary, more expensive
Pipe structure/bedding	Micro-deflections	Rigid pipes	Overall mechanical strength	Not affected by bedding condition, directly measures pipe's structural integrity	Rigid pipes only, not locates individual defects
	Natural vibrations	Empty sewers	Combined soil and pipe condition, crack region, exfiltration	Directly measures pipe's structural integrity without traveling entire length	Effect of bedding condition unknown, requires more sewer cleaning
	Impact echo	Large diameter sewers	Combined soil and pipe condition, crack region, exfiltration	Detects voids behind sewers, good for brick and concrete sewers	Manually operated equipment, does not locate individual defects
Bedding	Ground penetrating radar	Empty and partially filled pipes	Voids around pipe, water content in bedding	Detection of exfiltration, voids, rocks, and other objects in bedding	Interpretation of results, more expensive

flow within the pipe between two manholes). One vendor of this technology has suggested that the equipment may be used as part of a screening process to determine which sewer sections should be completely examined by mobile CCTV systems. The usefulness of stationary CCTV in this respect will depend on whether the damage that has been detected near a manhole in a sewer line is a representation of the defect across the entire section of the sewer line or not.

- **Mobile CCTV Systems**

Mobile CCTV systems are the most common means of inspecting sewer lines. This type of CCTV system uses a camera mounted on a robot that enters the sewer system. The camera lens generally faces the direction in which the robot system moves along the sewer axis, allowing the operator to examine and evaluate the entire length between a pair of manholes. It is possible to modify and customize this type of CCTV system to overcome many of the limitations of CCTV inspection discussed above. Some CCTV systems have 'pan and tilt' cameras attached to the robot, which can monitor defects that are overlooked by the front camera, thereby looking behind connections and other obstructions within the sewer line. Sonar or ultrasound systems are often attached to these robots to examine the sewer below the waterline. It is also possible to obtain CCTV equipment with a 'light line' attachment to assist in quantifying smaller sewer deformations. This system projects a line of light around the circumference of the sewer that is to be examined in order to assist in assessing the shape of the sewer.

1.3.2 Difficulties in CCTV Systems

Either form of CCTV inspection may overlook certain types of defects, especially those that are hidden from the camera view by obstructions as it looks down the sewer. Slight deformations of the sewer may go unnoticed, and any defect hidden beneath the water inside the sewer will definitely not be found.

It is a subjective and time-consuming task that considerably increases the inspection costs. Moreover, only gross anomalies are evident to the human eye, which reduces the detection of faults at early stages. Another drawback associated with these systems when used in such environments is the lack of visibility inside the pipes and the poor quality of the acquired images that hinders a complete assessment of the condition of the pipe and sometimes even the detection of large defects.

The described modifications to conventional CCTV systems can assist in interpreting the results of CCTV inspection, but are unlikely to be widely adopted unless their cost is not significantly higher than that of the conventional CCTV alone. Stationary CCTV cameras are therefore most likely to enter common use as they offer the opportunity to do preliminary examinations of the shorter pipes in a city-sewer system without cleaning facility. Mobile CCTV systems offer noticeable advantages in identifying the presence of deformation, but are merely an evolutionary enhancement of the standard CCTV system rather than a revolutionary improvement.

1.4 Sewer Pipe Condition Assessment Protocols

Sewer defect coding has gained paramount importance for the worldwide sewer rehabilitation industry to ascertain critical information regarding the subterranean pipe infrastructure [8]. The historical background of the development of sewer defect detection codes or condition assessment protocols dates back to 1977 when for the first time sewer defect detection codes were developed by Water Research Centre (WRC), UK. The first manual for sewer condition classification was published by WRC in 1980. On the basis of the guidelines provided by WRC, several condition assessment protocols have been developed throughout the world during the past twenty-five years.

The Japan Sewage Works Association (JSWA) [9] has classified three ranks (A, B, and C) for various faults in the sewer pipe system. Most visible faults are defined as rank A, visible faults as rank B, and hardly visible faults as rank C. Table 1.3 depicts the different fault classifications. Also, JSWA has defined the corresponding task to be done for each rank in Tables 1.3 and 1.4.

Table 1.3 Different fault classifications

Category	Rank A	Rank B	Rank C
Crack	The width of crack is more than 5 mm in the circumferential direction	The width of crack is more than 2 mm but less than 5 m in the circumferential direction	The width of crack is up to 2 mm in the circumferential direction
Water infiltration	Blown out	Blown out	Exuded
Root invasion	Invasion more than 50% of pipe	Invasion less than 50% of pipe	Invasion less than 20% of pipe
Pipe breakage	Heavy damage	Cracks throughout the length	Crack detection other than rank B
Joint and gap	Coming off entirely	Coming off partially	Small gap
Mounting pipe projection	Projection more than 50% of the diameter of the pipe	Projection between 25 and 50% of the diameter of the pipe	Projection less than 25% of the diameter of the pipe
Adhesion of mortar	Covers more than 30% of the diameter of the pipe	Covers between 10 and 30% of the diameter of the pipe	Covers less than 10% of the diameter of the pipe
Foreign substance	Blocked up with substances other than filth	Blocked up with obstacles and other substances	Blocked up with an obstacle itself such as a cement bag
Adhesion of grease	Block up more than 50% of the diameter of the pipe	Block up less than 50% of the diameter of the pipe	Block up less than 20% of the diameter of the pipe

Table 1.4 Corresponding task to be done for each rank

Levels	The inspection results	Corresponding tasks to be done
1	Too much rank A faults	Repair immediately
2	Too much rank B faults and a few rank A faults	Require simple maintenance; repairing can be extended by 5 years
3	Too much rank C faults, a few rank B faults, and no rank A faults	Require a simple maintenance, repairing can be extended to more than 5 years

1.5 Summary

With the aging problem as well as the increasing demand of reusing the damaged sewer pipes for laying fiber-optic cables, fast and accurate inspection of the pipelines is gaining importance at a rapid pace. Yet, at the same time due to the recent economic depression, the resource allocated for the sewer damage detection has been decreasing over the years. Despite such limitations, the urgent need for efficient inspection of the pipelines can no longer be ignored.

The deterioration process of sewers is highly complex and is dependent upon many factors. There are different types of sewer inspection techniques available, and the proper selection of these techniques depends upon the local requirements. Among these techniques, two types of CCTV inspection systems—stationary and mobile CCTVs—have been widely used for the investigation of sewer pipes. To resolve these issues economically and efficiently for monitoring and maintaining the sewage infrastructure, application of an automated fault detection system to the commonly used CCTV systems is considered to be highly beneficial.

Model Questions

1. Write briefly about the importance of sewerage pipelines and the need for its modernization.
2. Find out the advantages and disadvantages of automated inspection system over manual inspection system.
3. Write about various outcomes of faults in sewerage pipelines. Which faults are more critical and why?
4. What are different types of automated inspection systems used in sewerage lines?
5. What are various faults in pipes and how can one classify them?

References

1. http://www.jswa.jp/qa/4-2.html
2. http://www.jswa.jp/05_arekore/motto/07/index.html
3. Makar JM (1999) Diagnostic techniques for sewer systems. J Infrastruct Syst Nr 2 69–78

4. Kulandaivel G Sewer pipeline condition prediction using neural network models, Master's Thesis. Michigan State University, USA
5. Ahrary A (2007) Research on a vision system for autonomous inspection robots, Doctor Thesis. Kyushu Institute of Technology, Japan
6. Thornhill R, Wildbore P (2005) Sewer defect codes: origin and destination. U-tech underground construction paper
7. Japan Sewage Works Association, (2003) A sewer maintenance indication-first part. 124–125
8. Okahisa H (2009) Current situation and issues of Japanese sewer. Ministry of Land, Infrastructure, Transportation Tourism
9. http://www.jswa.jp/05_arekore/data-room/05/riyou/hikari.html

Chapter 2
Pipe Inspection Robots for Gas and Oil Pipelines

2.1 Introduction

Energy-related utilities such as fuel, gas, or oil supply lines and power generation sources such as nuclear and thermal power plants require an extensive network of pipelines for various transportation purposes. Figure 2.1a shows a typical network of pipelines generally used for oil and gas transportation.

These networks are in fact quite vast for any industrially developed country. The vastness of such operation in a developed country like USA is brought out in Fig. 2.1b.

Transportation of natural gas by pipelines is rapidly increasing in developing countries like India. Figure 2.1c shows the extent of Gas pipelines in India extracted from the International Energy Agency Report of 2010. Currently, India has about 11,000 km of gas pipelines. These pipelines, however, have limited life due to various types of static and dynamic loads originated from both inside and outside the pipe, as well as due to natural processes of degradation such as oxidation and corrosion of the pipe surface, and joint failures due to abrasion. The presence of internal defects/corrosion in a gas pipeline is generally detected by:

- Visual examination using inspection dig,
- External measurement such as electrical survey,
- Examination of corrosion coupons or probes placed inside the pipeline,
- Use of in-line inspection tool to identify areas of pitting or metal loss.

In this book, we will deal with various technologies developed specifically for in-line inspection.

H. Ogai and B. Bhattacharya, *Pipe Inspection Robots for Structural Health and Condition Monitoring*, Intelligent Systems, Control and Automation: Science and Engineering 89, https://doi.org/10.1007/978-81-322-3751-8_2

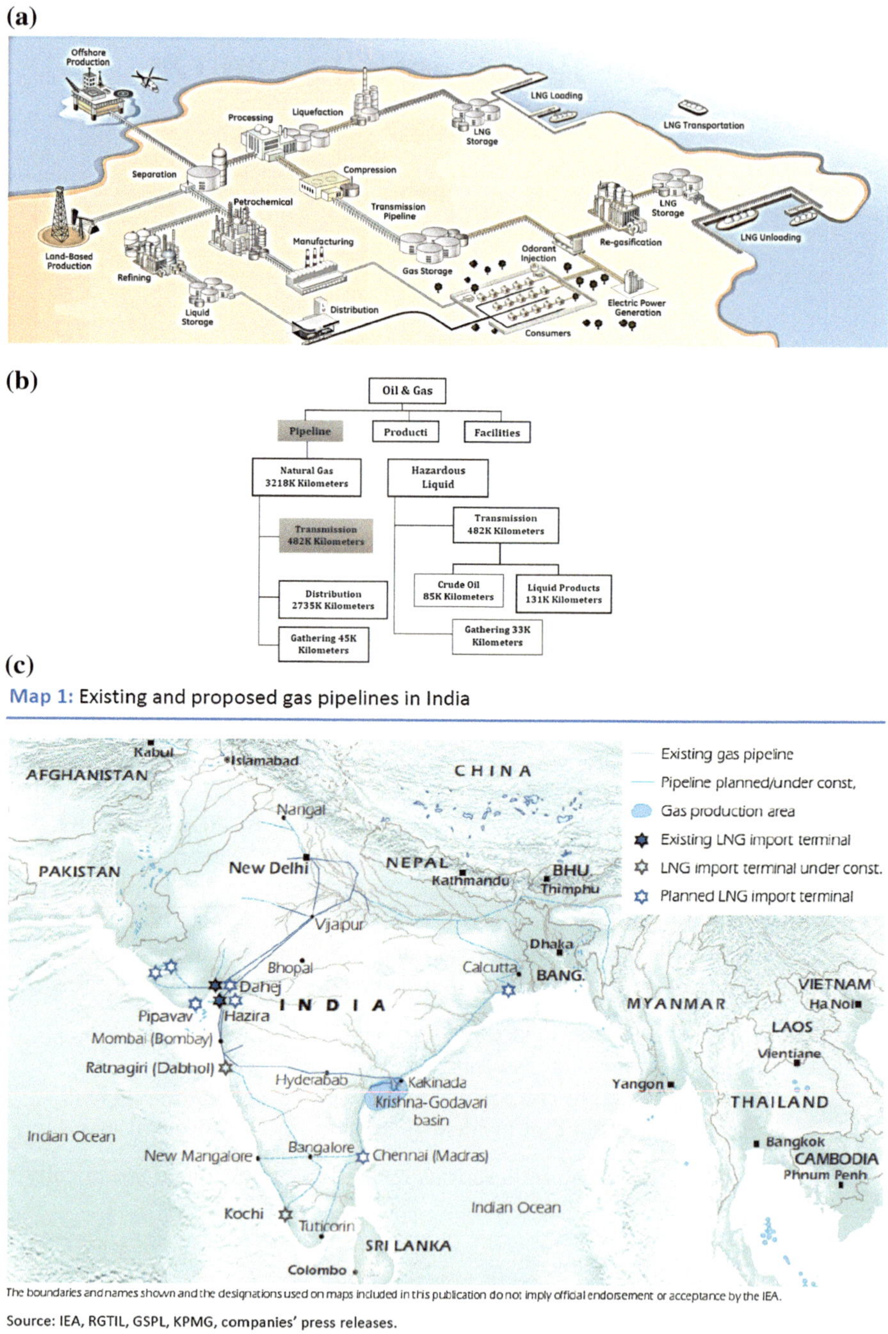

Fig. 2.1 **a** A typical pipeline network for oil and natural gas. **b** Oil and gas pipelines in USA (1998) [1]. **c** The extent of gas pipelines in India

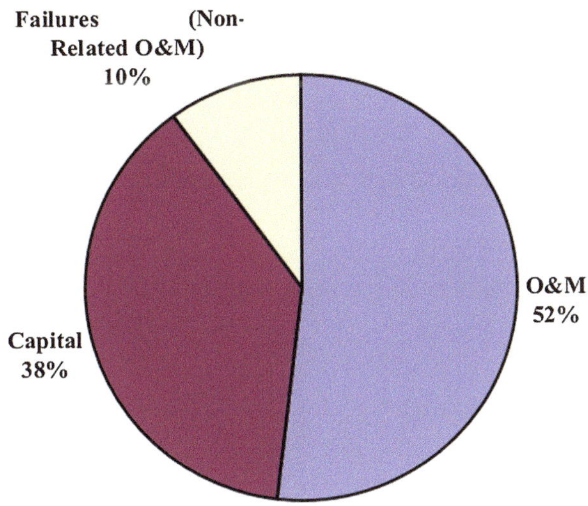

Fig. 2.2 Pattern of corrosion-related expenses in oil and gas pipelines [1]

The concept of structural health monitoring (SHM)[1] is immensely applicable in these fields to improve the life span of the pipelines and prevent catastrophic failures. As it is nearly impossible to equip the entire pipeline with distributed array of sensors, a better strategy that has been envisaged is to monitor the condition of the inner surface of the pipe with the help of a surveillance robot. A couple of decades ago, this technique was first introduced in nuclear power plants with the help of cable-drawn wheeled carts having a camera recorder at their apex. However, with the advent of sensors and actuators, fully autonomous robots based on various motion generation systems and integrated with multiple sensors are being developed today. Some of these robots travel hundreds of kilometers inside the pipelines to record the surface condition and location of damages. It may be understood from Fig. 2.2 that the operation and management of the oil and gas pipelines take up the major part of the expenses (more than 50%) in the transmission of oil and gas transport. Hence, a considerable research funding is allocated worldwide to find newer and cheaper ways of monitoring the pipelines to ensure safety and high performance of the system.

Such in-line monitoring system generally consists of two major subsystems—traction generation/locomotion system and damage sensing system.

The traction generation/locomotion system may be further classified into two broad categories: conventional locomotion and non-conventional locomotion. Conventional locomotion is further subdivided to three types:

(a) Wheel drive locomotion,
(b) tractor/truck-driven locomotion, and

[1]SHM is a complete process of design, development, and implementation of techniques for the detection, localization, and estimation of damages which is used in monitoring the integrity of structures and machines.

(c) Fluid-powered propulsion.

Non-conventional locomotion systems are more advantageous for very narrow (less than 100 mm diameter) and complex network of pipelines. Three such reported systems are as follows:

(a) Clamp and pull system similar to inchworm motion (speed 80 mm/s),
(b) Multi-legged system,
(c) Smart-ball system.

In the following sections, we will discuss various locomotion systems at length.

2.2 Traditional Locomotion Systems

Wheel drive locomotion system is one of the most commonly used techniques for driving pipe-crawlers. Figure 2.3a shows a typical wheel drive traction robot. These robots are generally used for large diameter pipes such as drainage pipes. They are also good for variable pipe diameters and high-speed driving. Maximum speed of such robots recorded in open literature is around 163 mm/s (approximately about 0.6 km/h!). In fact, for medium-sized pipes, the speed is even low about 30–80 mm/s. Of course, for effective SHM, forward speed is not of much concern as higher speed may compromise the accuracy of damage detection. A subgroup of wheeled robots is of the pressed-fit type (Fig. 2.3b) and is especially suitable for curved pipes and vertical climbing.

For uneven and greasy or muddy pipe surfaces, often better grips are required. In such cases, tractor/truck-driven systems are more useful. Figure 2.4 shows a typical belt-driven system. Such robots are also resource-effective from the manufacturing point of view as they are not only simple in design but also have a better load-carrying capacity than wheel drive systems. However, these robots are generally slower than the wheel-driven robots (recorded max speed is around 40 mm/s).

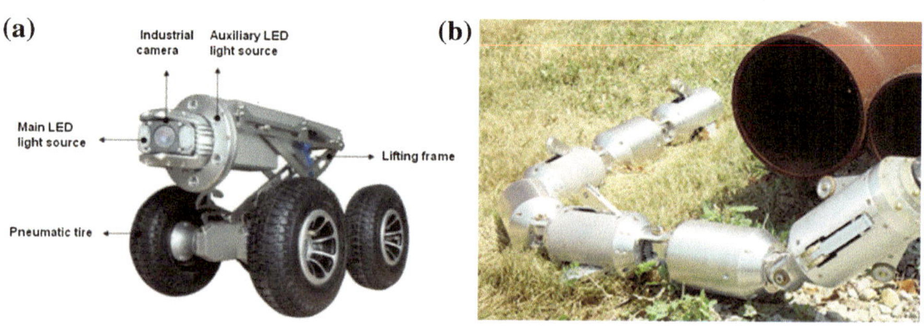

Fig. 2.3 **a** A typical wheel-type pipe inspection robot with an onboard camera system, **b** a pressed-fit wheeled robot system [2]

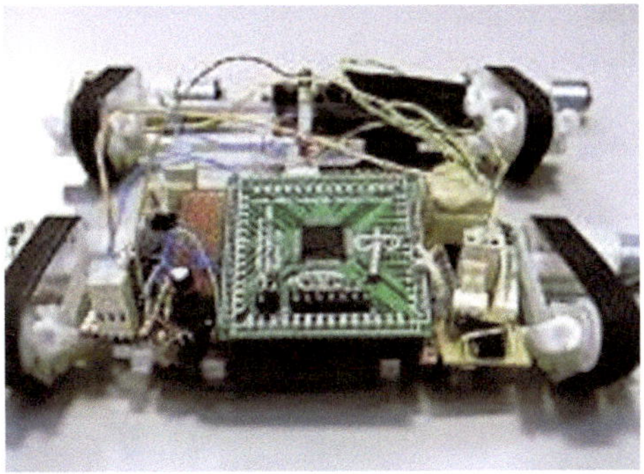

Fig. 2.4 Belt-based crawler robot [3]

Propulsion-driven systems are generally used in relatively large diameter water and gas pipelines. For such applications, passive robots are developed that can mostly extract the flow energy for locomotion. Pipe Investigation Gauge (PIG) as shown in Fig. 2.5 is one such example. The weight of each module of the robot is limited by the kinetic energy available from the flow.

2.3 Non-conventional Locomotion

Clamp and pull systems are generally used in smaller pipelines where the space is limited and the flow energy is not sufficient to pull the robot. This system is essentially very similar to the inchworm movement which generates motion by a sequence of rear clamping–expanding and front clamping–pulling action. Figure 2.6a shows a typical miniaturized clamp and pull robot. There are numerous systems designed to accomplish this motion. However, the basic concept remains the same. Figure 2.6b shows the five-stage sequence of inchworm motion in a SMA-activated modular system.

There are a few other exotic varieties of the motion generation system such as walking robots equipped with legged motion. These types of robots are specially

Fig. 2.5 A propulsion-driven PIG system with magnetic flux sensors

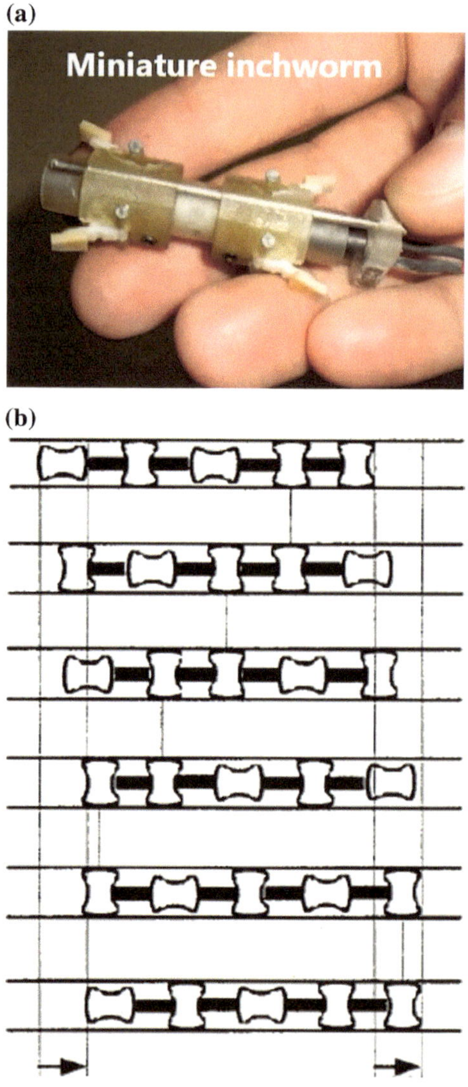

Fig. 2.6 **a** A miniature inchworm robot developed at Technion [4]. **b** A five-stage inchworm motion based on SMA

suited for uneven surfaces and pipes with complex connections. A few of these legged variations of walking robots are shown in Fig. 2.7a–c.

The legs are used to press against the pipe wall to provide enough traction for the robot to climb or walk inside a pipe or duct. Unlike most of the visual feedback-based robots, movement of this robot is based on tactile sensing at the feet and contact sensing along the legs. The control of the robot is generally divided into three control units: the main control unit, the leg control unit, and the force/velocity control unit. During motion, at a very local level, each individual leg lifts itself and tries to judge the stability of the robot. If the stability is compromised during several

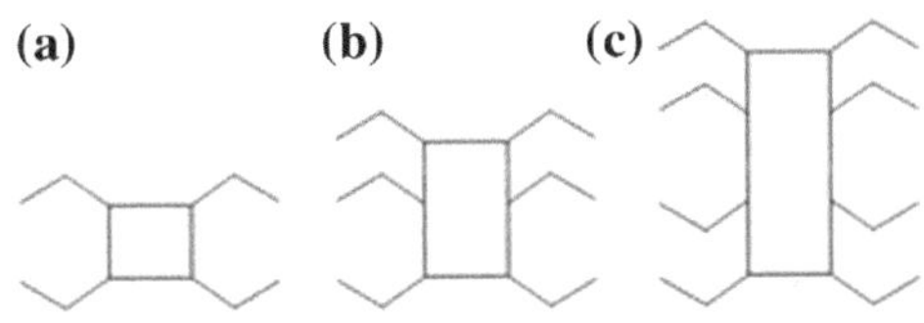

Fig. 2.7 **a–c** 4, 6, and 8 legged robots

trials, then the main control unit is informed about the change of stance. It is found that articulated leg-type robot with such reflexive behavior can easily pass through obstacles such as bends, diameter variations. However, generation of power and manufacturing complexity of the leg system are two of the major hindrances to the development of this technology.

Smart-ball leak detection [5] is a free-flowing system used to locate leaks and gas pockets in pressurized pipelines (see Fig. 2.8). It can complete long leak detection surveys in a single deployment without disrupting regular pipeline service. The tool is equipped with highly sensitive acoustic sensor that is able to locate 'pinhole'-sized leaks. The Smart-ball is inserted into a pipeline and travels with the water flow for up to 12 h while collecting information about leaks and gas pockets. It requires only two access points, one for insertion and the other for extraction, and is tracked throughout the inspection at predetermined fixed locations on the pipeline.

The detection system consists of an inner aluminum alloy core containing an acoustic sensor and circuitry. The aluminum core is in turn encapsulated inside a foam ball. The foam ball provides the appropriate mass (size and overall weight) that allows the device to be propelled by the water flow. It also absorbs any noise that the device may make as it traverses through the pipeline. While the ball is traversing inside the pipeline, a transponder within the Smart-ball core emits high-frequency, timed acoustic signals that are detected by Smart-ball Receivers (SBRs) on the pipe surface. The Smart-ball Receivers track the Smart-ball's movement and location, correlating its position at any time with reference to acoustic events recorded on the acoustic sensor contained within the Smart-ball. Once the Smart-ball has traversed the entire length of the pipeline, it is typically captured and retrieved either in a specially engineered net or in an open channel. The recorded data are subsequently evaluated to determine the presence and location of leaks or pockets of trapped gas. Smart-ball requires a minimum of 100 mm flange opening with a full port valve for insertion into the pipeline. Once deployed, it can move through in-line valves, reducers, and other fittings, as well as navigate through turns and profile changes.

In the following section, we will have an elaborate discussion of a few more complex traction mechanisms.

2.4 More Complex Traction Systems

In this section, we will discuss some complex locomotion systems which are primarily based on the conventional locomotion systems described earlier. These are, however, more complex in nature since in addition to simple translation motion, they have to perform additional functions. Out of numerous multi-function locomotion systems, the following four are found to have many industrial applications:

(i) Adaptive system—to move inside variable diameter pipes,
(ii) Active steering system—that can take care of bends and branches,
(iii) Self-retrieval and obstruction mitigation system and finally,
(iv) Vertical climbing system.

In what follows is a brief description of some of the multi-functional robots developed at various robotics laboratories.

2.4.1 Three-Wheeled Self-adjusting Vehicle for Variable Pipe Diameter

A three-wheeled robot (FERRET-1), [6] whose shape looks like the alphabet '**A**' is shown in Fig. 2.9. At the end of the two legs, spherical bearings are installed, and depending on whether the vehicle type is tractive or non-tractive, a driving wheel or another spherical bearing is installed at the vertex joining the two legs. The two legs are connected by a spring between them. The purpose of the spring is to provide enough tractive force on the wheels so that the wheels do not slip. Two different types of driving wheels are considered in this design which can guide the vehicle posture to the maximum diameter of the pipe.

Fig. 2.8 Smart-ball Technology [5]

2.4.2 *MOGRER—A More Complex Variant of the FERRET Robot*

In a more complex form, another design of the three-wheeled pipe-crawling robot with two arms and two links as illustrated in Fig. 2.10a–b is developed [7]. The shape of this robot resembles the shape of a pair of scissors. This scissor shape helps it to self-adjust the height of the robot when it passes through pipes with varying cross-sectional diameters. In a modified version, levers are used at a fixed angle from the arm to connect arms with the spring (as shown in Fig. 2.10b). It is observed that the lever-based system works more effectively for the generation of traction force.

2.4.3 *Differential Drive In-Pipe Robot for Moving Inside Urban Gas Pipelines*

Figure 2.11 shows the concept of a differential drive pipe-crawling robot. In this model, three similar kinds of wheel assemblies, circumferentially 120° apart from each other, are used in the robot. The height of the wheel set can be varied by adjusting individual springs for each wheel. The springs are used to maintain traction force on the wheels and also to self-adjust the wheels to pass through the varying diameters of the pipelines.

2.4.4 *Robot with Active Steering Capability for Urban Gas Pipelines*

Figure 2.12 shows the outline of the steering system of a modular robot called MRINSPECT [8] which is used to carry out inspections inside pipes of diameter 150–200 mm. The entire robot is divided into three modules—driving module, control module, and inspection module. All these modules are connected by

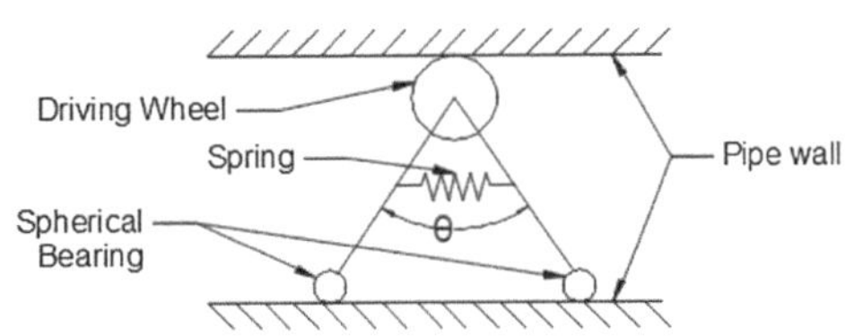

Fig. 2.9 A three-wheeled FERRET–1 robot

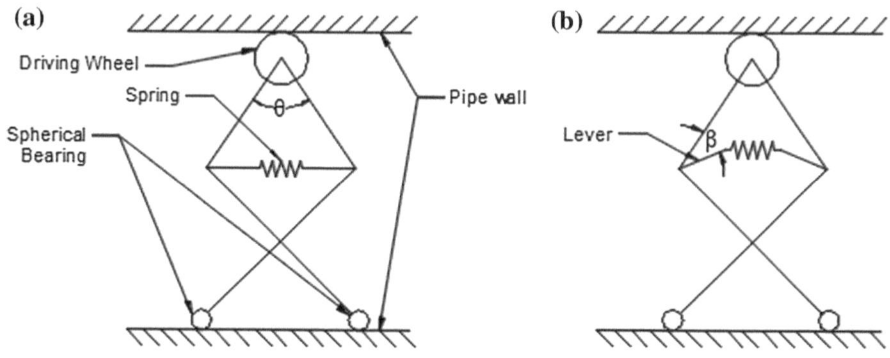

Fig. 2.10 Multi-link scissor-like robot with height control mechanism

Fig. 2.11 Adaptive differential drive system

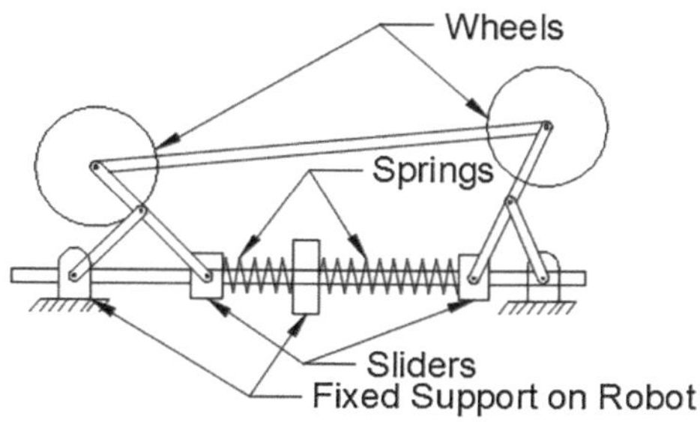

universal joints. Two sets of driving modules are installed in the front and the rear end of the robot, so that both forward motion and backward motion can be achieved. Each driving module is further comprised of two parts where each part is connected by a double active universal joint. It is specifically used to steer the front part of the driving module through bends and fittings. Active driving force is provided by DC motors, worm gears, and timing belts on the rear part of the driving module. Figure 2.12 explains the pantograph-type leg mechanism for the driving modules. In each part of the driving module, three sets of these mechanisms are installed. A spring is used to provide enough traction force to the robot to crawl through the pipe. Also, it helps to adjust the leg height when the robot passes through the varying diameters of a pipeline. A charge-coupled device (CCD)-based camera is installed on each driving module to record the movement of robot inside the pipe and also to guide the controller in terms of steering the vehicle. Inspection modules are considered optional and can be replaced by an appropriate sensor module as and when required.

2.4.5 A Novel Turbine-Propelled Self-drive Pipe-Crawling Robot

Figure 2.13 describes the design of a passive pipe-crawling robot, which uses the flow of fluids to traverse inside a pipe [9]. One of the key features of the robot is its inclined bristles. These bristles help the robot to traverse in only one direction. Further, a nut, a double screw thread, and a turbine are used to transform the rotational motion of the turbine to translatory motion of the robot. The speed of the robot depends on the pitch of the thread which can be adjusted.

2.4.6 Robot with Active Pipe Diameter Adaptability and Automatic Tractive Force Adjustment

Figure 2.14 shows a novel technique to maintain the tractive force between the pipe wall and the wheels, using parallelogram wheeled leg mechanism, adjusting motor, ball screw, nut, and pressure sensor [10]. Three such assemblies are used which are 120° apart circumferentially. This robot is designed to adjust itself automatically between pipe diameters ranging from 400 to 650 mm; whenever the robot experiences differential diameter, the tractive force between the pipe wall and the wheels changes, which is measured by the pressure sensor located in between the sliding bush and the nut. This change in pressure triggers the adjusting motor to rotate clockwise or anticlockwise according to the requirement, which in turn helps it to maintain the tractive force between the pipe wall and the wheel. The same technique is used to deal with the bends in the pipelines.

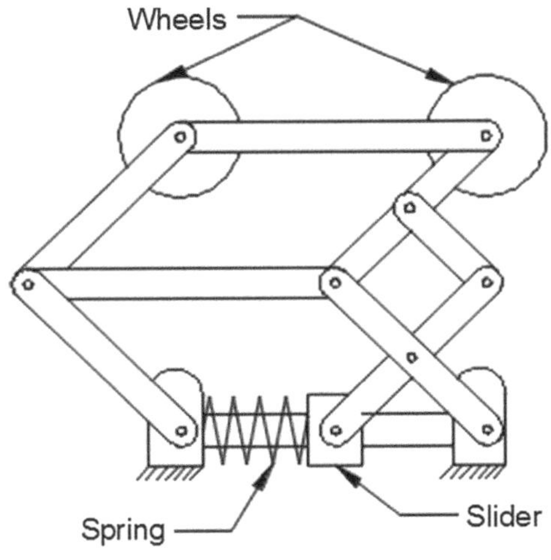

Fig. 2.12 A robot with active steering

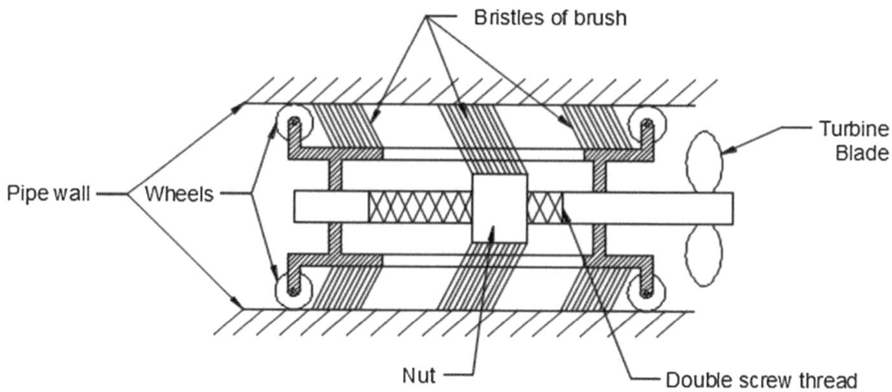

Fig. 2.13 An energy-harvesting passive pipe-crawling robot [9]

2.4.7 *Adaptive Mobile Robot for In-Pipe Inspection Task*

Figure 2.15 shows the schematic of another novel pipe-crawling robot which is both adaptive in size and capable of clearing obstructions inside a pipe. The robot consists of six sets of such mechanisms 60° apart. Each mechanism consists of three sets of wheels. The first two sets of wheels are termed as rotor outputs 1 and 2, respectively. All wheels in rotor output 1 are inclined with respect to the z-axis and fixed to the rotor shaft. This is provided with an intention to achieve spiral movement for the robot as the rotor shaft is rotated which is useful for continuous monitoring. Rotor output 2 wheels are free to rotate about the z-axis and get engaged to rotor shaft only when rotor output 1 wheels get stuck. This is achieved by using a set of sun and planet gear combination. Whenever wheels linked to rotor output 1 get locked, rotor output 2 provides extra power to the robot to overcome obstacles. The rear sets of wheels are attached to stator part of the robot on which the actuator is mounted. This novel system also reduces the number of actuators (from 3 actuators to 1 actuator) required to drive a robot inside a pipe. Hence, the power consumption of the robot reduces significantly.

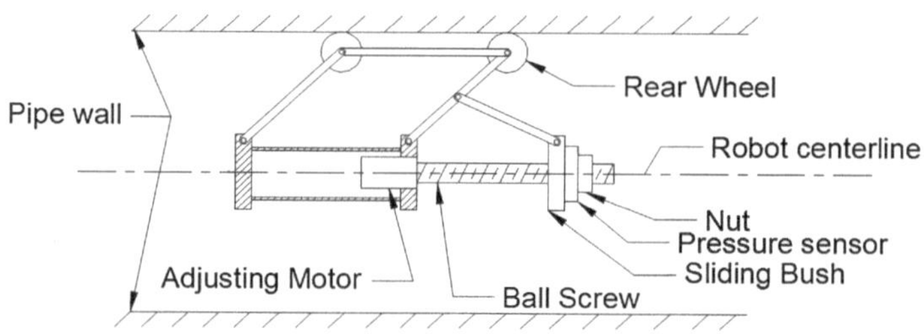

Fig. 2.14 A PCR with adaptive shape and tractive force generation system [11]

2.4.8 Adaptive Track-Based Robot for In-Pipe Inspection

The advantage of track-based locomotion is already discussed in Sect. 2.2. A modified track-based pipe-crawling robot is shown in Fig. 2.16. The robot consists of three modules: track driving module, center module, and pantograph connecting module. The track module consists of a frontal track and two rear tracks. Frontal and rear tracks are connected by a compliant active joint. The track module is attached to the center module by pantograph connector. The pantograph connector module also helps the robot to adapt for variable pipe diameters. Three sets of combined pantograph connecting module and track module are attached to the center module at 120° apart. One servomotor is used on rear track of each set of track to provide driving torque to the robot. Transfer of drive torque from rear to front track is achieved by a set of gears. By rotating the lead screw in clockwise or anticlockwise direction, the legs of pantograph can be brought closer or apart, which in turns helps the robot to adapt inside varying cross section of pipes.

2.4.9 Micro-In-Pipe Robot Based on Shape Memory Alloy Actuator

Figure 2.17 shows a shape memory alloy (SMA) spring-based mini-pipe-crawling robot which can crawl inside pipes of diameter varying from 50 to 80 mm [13]. The robot is divided into two identical parts called gripping module and driving module. The gripping and driving modules are connected using four SMA springs. During the motion of robot, only one module is engaged to the pipe and other move by compressing or expanding the SMA springs. Each gripper consists of friction spring strips, an electromotor, and a strap. The electromotor is used to wind or unwind the strap which in turn changes the distance between lower and upper cover of the gripper. This results in change in curvature of friction spring strips which causes the gripper to grip or release. Four two-way memory effect SMA springs are used to connect the two grippers. These springs can compress or expand to bring the grippers closer or apart. Compression and expansion of springs are controlled by varying the temperature of the springs, which is achieved by passing current

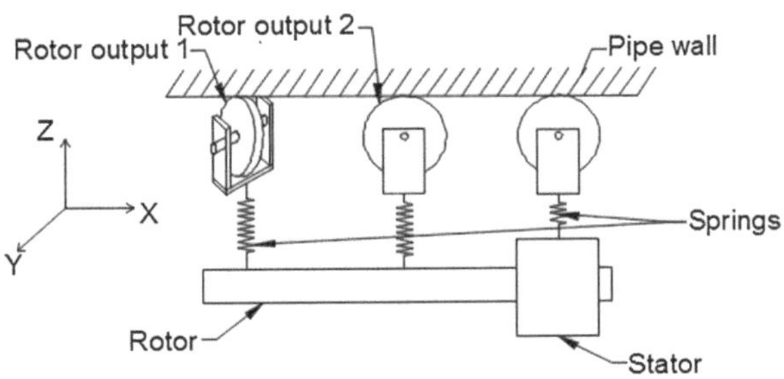

Fig. 2.15 Adaptive, obstacle clearance system

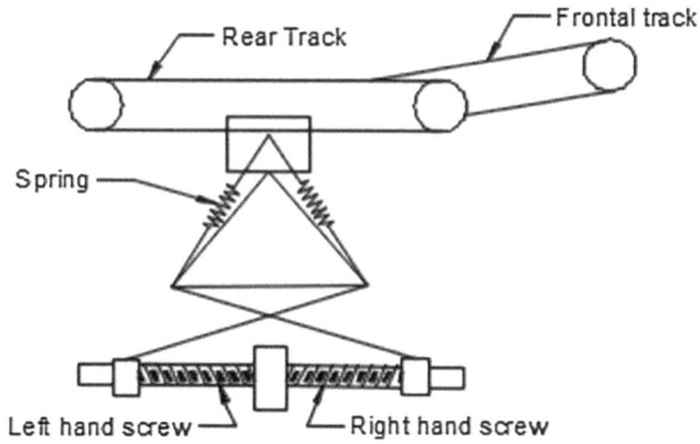

Fig. 2.16 A track-based adaptive pipe-crawling robot [12]

through the spring wires. For steering, the robot springs are selectively compressed. These types of robot are energy efficient and compact. Hence, such robots are effective for pipes of lower diameter. A speed close to 3 mm/s is reported to be achieved in this system.

2.4.10 Multi-functional Pipe Investigation Robot (PIR) with Linkage-Type Mechanical Clutch

Figure 2.18 shows a linkage-type mechanical clutch system [14], which helps the vehicle to operate in two different modes: locomotion mode and retrieval mode. Drive wheels are used for forward and backward locomotion of the vehicle. The rear and clutch wheels are idle wheels. There are two sets of mechanisms highlighted in Fig. 2.18. The mechanism with red border is basically a five-bar linkage mechanism, which is used to adjust the height of the vehicle and keep the line joining centers of rear idle wheel and drive wheel parallel to the axis of the vehicle. The mechanism highlighted in blue border is a four bar mechanism and is used to adjust the height of the clutch wheel. Once, height of the set of drive wheel mechanism is lowered, due to the presence of four bar mechanism in clutch wheel, its height raises. Hence, when the drive wheel set disengages from pipe wall, the clutch wheel system gets engaged, making it easier to drive the vehicle out of pipe by simply pulling it from the back.

2.4.11 Planar Multi-functional PIR with Two Sets of Wheels

Figure 2.19 shows a drive mechanism based on only two sets of wheel assembly (180° apart). Using two sets instead of three provides more space to mount sensors

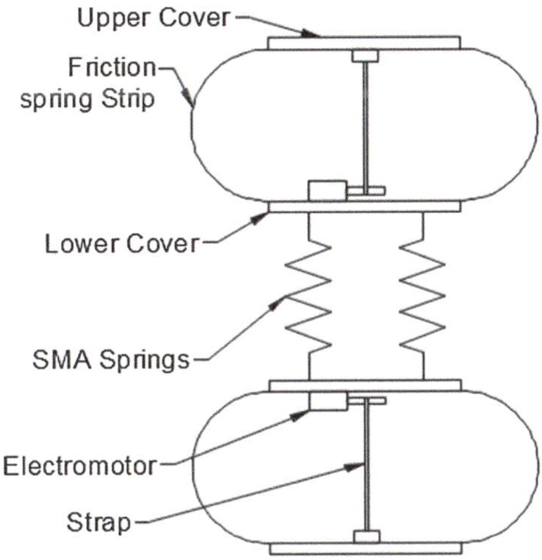

Fig. 2.17 A shape memory alloy-based micro-adaptive pipe-crawler

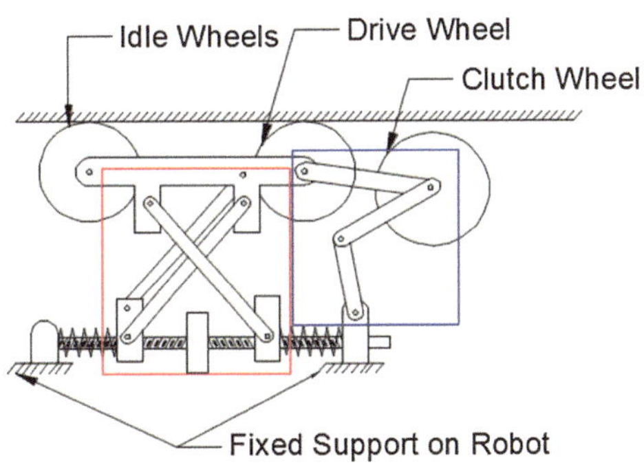

Fig. 2.18 A multi-functional PIR

and also reduces design complexities. Each set of wheel assembly consists of a pair of wheels where the front wheels are used to steer the robot and rear wheels are used for driving the robot. This small robot is developed to travel inside a medium-sized pipe of internal diameter 100 mm. Depending on the direction of steering wheel, the robot can travel in a helical or straight path. Steering wheels are also used to turn the robot in elbow or T-joints. Working of this robot is categorized into three parts, i.e., driving, detecting, and searching mode. The robot is equipped with three cameras, one at the front and two at the sides. In the detecting mode, the front camera locates the defect, while to collect detailed information of defect the two sidewise cameras are used.

2.4.12 *Planar Adaptive Mechanism for an In-Pipe Robot*

Another novel planer mechanism, consisting four wheels, is described in Fig. 2.20. Four external arms are hinged to the vehicle body and the wheels are attached to the arms by a U-frame arrangement. It is also referred as adaptable quad arm mechanism. Here, all four wheels are actively driven by separate motors. By controlling the direction of rotation of wheels, the robot can be made to move forward/backward or the horizontal distance of the wheels can be varied. This mechanism allows the robot to travel through bends of zero radius of curvature and bends of varying diameter. Since each wheel can steer with respect to arms, the mechanism is termed as swivel hand mechanism. This mechanism allows the robot to drive in a spiral motion and also helps to orient inside the pipe.

2.4.13 *Peristaltic Pipe-Crawling Robot for Long Distance Inspection of Sewer Pipes*

A novel peristaltic pipe-crawling robot is designed and fabricated for the inspection of small pipelines (<70 mm diameter) [15]. Locomotion of this robot is inspired from the movements of earthworm. The important member responsible for such movement is the artificial muscle, which is made up of micro-carbon fiber and low ammonia natural rubber latex. Figure 2.21a explains the arrangement of fibers and latex of artificial muscle. Driving units are actuated by compressed air. Components of a unit of this robot are described in Fig. 2.21b. The entire robot consists of six such units, and a camera is mounted on the front unit for inspection of pipeline. Electrically controlled valves are used to inflate or deflate artificial muscle. Since carbon fibers do not elongate, the rear and front part of a module comes closer when a module is inflated and vise versa. This helps the modules to grip the pipe wall and move the robot forward.

Fig. 2.19 A planar PHMR

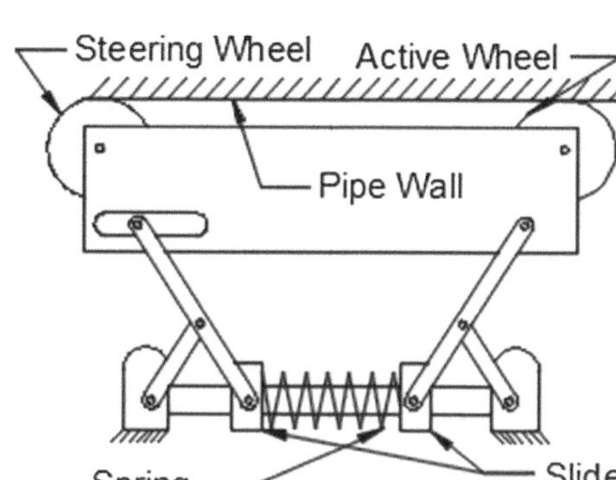

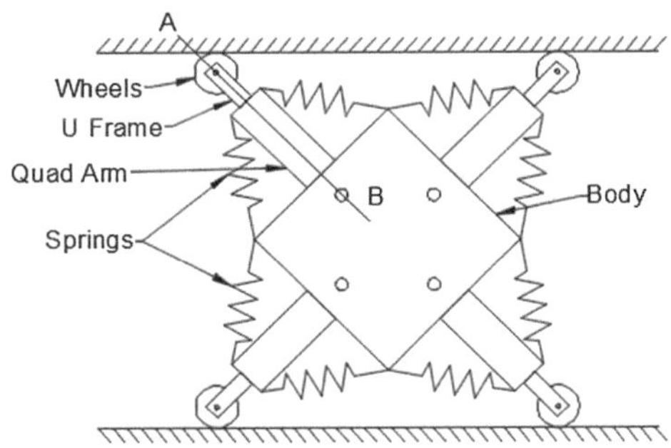

Fig. 2.20 Another planar mechanism with adaptive locomotion capability

2.4.14 Inchworm-Based Vertical Climbing Robot with Adaptive Size

Figure 2.22 shows the design of a robot whose crawling mechanism is based on an inchworm. The robot consists of two clamping devices, two universal joints, and an expansion device. Compressed air is used as a source of energy for driving the robot. The clamping device is based on a four bar mechanism. Figure 2.22 shows the assembly of clamping device. Three sets of this mechanism are mounted on each set of clamping device. The clamping devices produce enough traction force for the robot to enable it to climb along a vertical pipe carrying its weight along with tether cables and sensors. This robot has been shown to traverse inside a pipe of diameter varying from 100 to 300 mm.

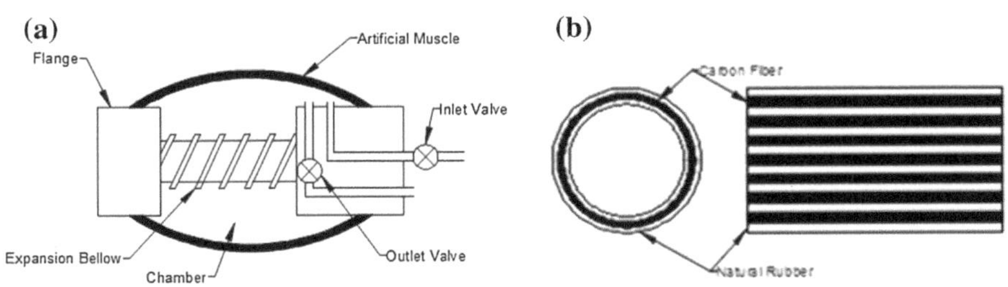

Fig. 2.21 A peristaltic PHMR: **a** transverse cross-sectional view and **b** longitudinal cross-sectional view

2.5 Various Types of Wheel Assembly for the Pipe-Crawling Robots

Wheels can be viewed as the dynamic interfaces between the health monitoring system and the inner surface of the pipeline. Similar to the ground vehicles, a good design of wheel assembly ensures less vibration and noisy operation of the robot. In this section, we will take a closer look on the wheel assembly of various types available for the crawler robots. The entire section is divided into three subsections: First, we will briefly discuss the major challenges faced during the inspection process; next, we will deal with various types of wheel assemblies; and finally, we will present a comparison table which shows the cumulative advantages and disadvantages of a few wheel assembly systems.

2.5.1 General Challenges Faced by the Pipe-Crawlers

The following are the major challenges faced by the pipe-crawlers inside the pipelines:

- Wear and tear due to long distance travel inside the pipelines,
- Continuous position monitoring of the robot for location indication,
- Slippery inner walls of the pipeline causing unstable motion,
- Turns and bends in the pipeline,
- Non-uniform gas flow,
- Turbulence of the gas flow, and
- The presence of corrosive agents.

In what follows are brief descriptions of various wheel assembly systems designed for this purpose.

2.5.1.1 Single-Body Three-Wheeled System

A simple three-wheeled model generally consists of wheels connected with springs attached to the main body, to accommodate any small changes in the pipe diameter (Fig. 2.23). However, this simple arrangement is often not enough to adapt with the large changes in the pipe diameter. Therefore, the basic model has to be corrected and modified to face these challenges. There has to be a mechanism, which can move the whole frame of the robot so as to change the radius of the robot.

The movement of the robot can be carried out with the help of an adaptive mechanism by actuating the pipe diameter adjustment of the tractive force. A stepper motor is used for pushing the spring, which in turn causes the increase in

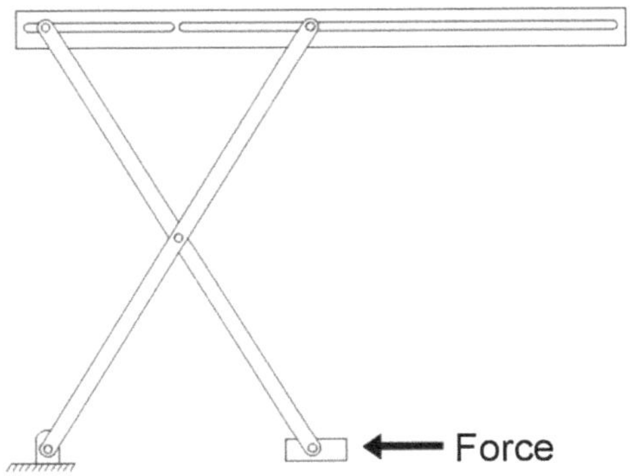

Fig. 2.22 Inchworm-based crawling mechanism

the tractive force exerted on the walls of the pipe. The inspection robot has to change its diameter in order to mitigate various obstacles such as changing pipe diameter, bends, curves.

2.5.1.2 Multi-sectional Wheeled Robot

This type of wheel system ensures active steering capability and diameter adaptability. The robot is designed as a serpentine structure (Fig. 2.24) [16]. Two active driving vehicles are located in the front and rear of the system, respectively, which provides sufficient traction for the entire robot. In this robot, the steering mechanism is based on double active universal joint, which provides omnidirectional steering capability. This system helps to mitigate challenges/defects such as branching,

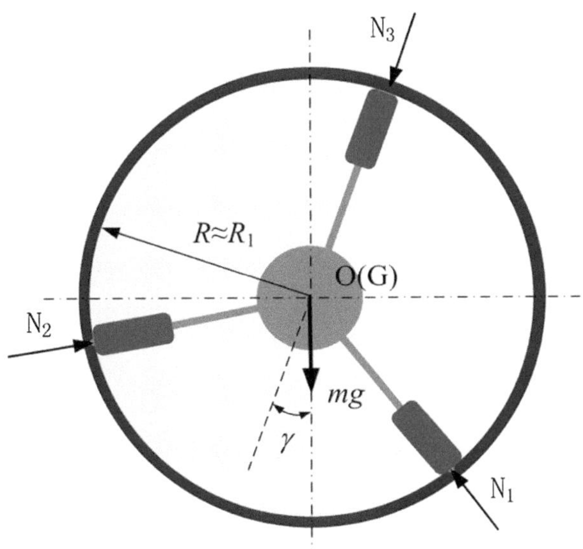

Fig. 2.23 Front view of a single-body three-wheeled crawler system

Fig. 2.24 View of a multi-sectional wheeled robot

passing through the reducers, and valves with mechanical damages such as dents, gouges.

Various segments of the driving module are connected with universal joints. Both the front and rear segments have three-wheeled leg mechanisms. The three-wheeled mechanism continuously expands and contracts according to the terrain of the inner walls of the pipes. In this robot, the traction force is mainly created by the wheels on the legs which are actuated by a DC motor in the rear segment. The power of the motor is evenly transmitted via worm gear mechanism.

2.5.1.3 Robot with Annular Wheeled System

This system mainly consists of two parts—the stator and the rotor, connected by an active joint including a DC motor with a reducer and, in some cases, a universal joint [16]. The stator has a set of wheels which provides motion along the axis of the pipe, whereas the rotor is equipped with wheels tilted at a small angle with respect to the plane perpendicular to the axis of motion. This setup ensures the constraining of the stator so that while it moves along the tube axis, the wheels of the rotor move along a helical path (Fig. 2.25).

The wheels on the stator and the rotor are present to guarantee stability against overturning. Universal joints are provided for locomotion if there are small diameter pipes with abrupt changes in the angles of the pipes.

2.5.1.4 Wheelless Pipeline Inspection and Condition Analysis System

This system is generally used for small diameter water pipe inspection (Fig. 2.26) [17]. It uses remote field technology (RFT), in which no contact is required with the walls. It is insensitive to the presence of boundary walls and therefore suitable for thinning detection. It is waterproof and can measure through wax or sand and

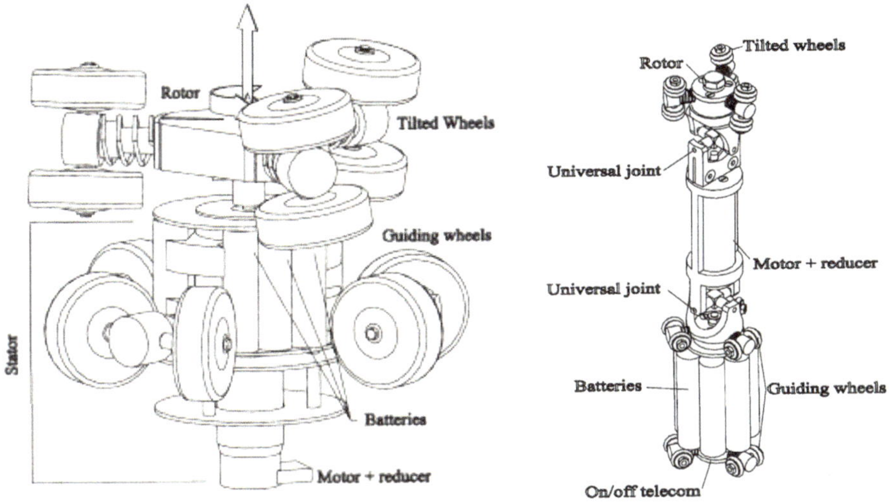

Fig. 2.25 Robot with annular wheeled system

Fig. 2.26 A wheelless pipeline inspection system

magnetic liners. This robot is suited for both cast and ductile iron pipelines. It needs a very small pushing by product flow/air pressure (about 2 bar). It is reported to have a speed of around 1–2 km/h.

2.5.2 Performance Summary

A brief performance summary of the four systems discussed above is provided in Table 2.1.

Table 2.1 Performance comparison of various drive systems

Advantages	Disadvantages
Model-1 (ref. 5.2.2.1)	
Advantages • Low weight • Simple construction	Disadvantages • Poor diameter adaptability • External power is required
Model-2 (ref. 5.2.2.2)	
Advantages • Easy steering • Diameter adaptability • Modular system	Disadvantages • Heavier system • Low speed due to poor friction • External power required
Model-3 (ref. 5.2.2.3)	
Advantages • Compact and lightweight • Diameter adaptability • Good friction characteristics	Disadvantages • External power required • Complex system
Model-4 (ref. 5.2.2.4)	
Advantages • Lightweight • No requirement of external power • Diameter adaptability • Easy steering	Disadvantages • Needs enough pushing force • Less sensitive for large diameter pipes • No speed control. Braking system

2.6 Various Sensing Mechanisms Used in Pipe-Crawlers

A functional and autonomous robot for SHM needs to carry with it an array of sensors, data recorder, communication system, power supply, etc. However, a single robotic unit cannot carry all the instruments as the load-carrying capacity of any robot is quite limited either due to the limitation of propulsive force or due to the limited power supply. This is the reason why such robotic systems are generally modular in nature. A typical robot consists of four types of modules, i.e., tractive vehicle, ultrasonic testing vehicle, control vehicle, and coupler module. In last few sections, we had focused on various technologies applied for the locomotion of the robot. In this section, we will discuss the sensing systems applied for the SHM of pipe surface.

Typically, for sensing of defects in gas pipes, the following systems are used:

(a) Optical Sensing

- Visual inspection-based sensing,
- Laser profiling,
- Microwave sensing.

(b) Vibration and Sound Wave-based Sensing

- Sonar profiling,
- Ultrasonic sensing,
- PVDF-based vibration sensing.

(c) Indirect Sensing

- Magnetic flux leakage sensing,
- Villary sensing,
- Gas sensing.

In the following section, we will provide brief description of some of these sensors.

2.6.1 *Visual Inspection of Pipe Health Conditions*

The biggest challenge of optical scanning in pipelines is the absence of proper illumination. As shown in Fig. 2.27, servo-controlled cameras with night mode are generally used for visual inspection of pipe surfaces. These cameras provide the ability to have a real-time view of the piping system and are able to record both images and videos. Camera tractors are used frequently in sewer pipe health monitoring. The crawler moves inside the pipeline by using two sets of belt-driven systems. An external power cable is attached to provide power for motion and camera. A major challenge in this type of sensing system is the selection of proper camera lenses. Most of the lenses being fish-eye-type provide unnecessary details of the open part of the pipe and less details of the pipe surface. Several omnidirectional cameras with cylindrical strip vision have been recently designed to overcome this problem.

2.6.2 *Laser Profiling Sensor*

This type of sensing is used for indicating deformations and corrosion inside the pipeline with high precision and for a broad range of diameters. The laser profiling is reported to measure the diameter and concentricity of the pipe ranging from 100 mm to 4 m diameter, performing up to 250,000 measurements per min with

Fig. 2.27 Visual inspection system

Fig. 2.28 A typical laser profiler

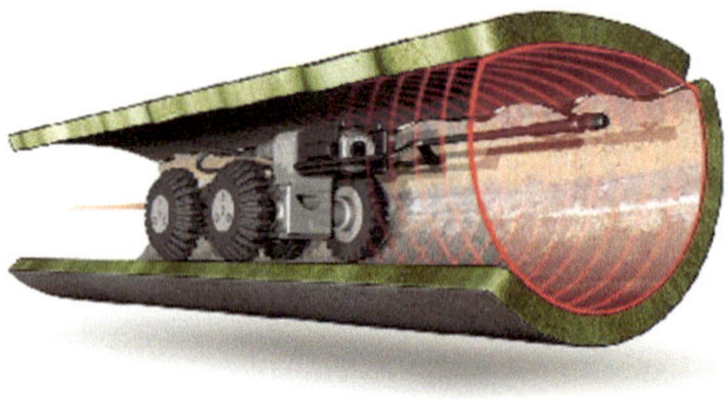

±1% accuracy! The output data are generally collected over the entire length of the pipe and can be used to generate statistical reports, as well as 3D visualization of the interior wall of the pipe. For an effective 3D visualization, however, multiple lasers are used that can measure deformations based on optical triangulation. In some literature, the use of rotating multiple lasers is proposed which along with forward-moving probes can generate high-precision mapping of the pipe surface (Fig. 2.28).

2.6.3 Sonar Pipe Profiling

Sonar sensors are mainly used for underwater pipe profiling where CCTV cameras are not useful, particularly in pipes carrying murky fluids. Often, it is also used along with laser profilers. This system utilizes high-resolution and short-range sonar for real-time cross-sectional views of the pipe. This type of sensor can be used in submerged and semi-submerged pipelines ranging from 200 mm and above. The system consists of piezoelectric transducers which generate a characteristic acoustic signal and measure the reflection of the same from the pipe surface. It provides accurate dimensional data on grease accumulation, blockages, pipe deformation, etc., thereby providing a 2D profile of the interior of the pipe wall (Fig. 2.29).

2.6.4 Polyvinylidene Difluoride (PVDF) Sensor

PVDF is a relatively low-cost polymeric piezo-sensor which generates output voltage in the range of few volts when the sensor probes scan the pipe inner surface

Fig. 2.29 A typical sonar profiler

Fig. 2.30 A PVDF sensor-based robot

(see Fig. 2.30) [18]. These types of sensors are usually supplied in the form of thin film, typically ranging from 9 to 110 μm thicknesses. Specially designed cantilever beam surface coated with PVDF films is used as rotating sensory probes. The electrical charge generated in the PVDF strip is proportional to the change in the mechanical stress. Hence, this sensor can be used to detect finite projections in the pipe surface. One of the attractive features of this sensor is that it is much less data-intensive and hence can be used for very long-range scanning applications.

2.6.5 LIDAR (Light Detection and Ranging)

LIDAR is a remote monitoring technology that basically measures the distance of a target by illuminating it with a laser and analyzing the reflected light. LIDAR sensing uses ultraviolet, visible, or near-infrared light to image targets and can be used with a wide range of objects, including nonmetallic objects, chemical compounds, clouds, rocks, rain, and even single molecules. Airborne LIDAR Pipeline Inspection Service is based on a mid-infrared Differential Absorption LIDAR (DIAL) chemical sensor. The principle of DIAL is based upon the selective absorption of laser light by different chemicals. During operation, the laser beam is transmitted down from the aircraft to illuminate the area on the ground above and around the buried pipeline. The light reflected from the ground is collected by the receiver of the sensor, and the amount of reflected energy is thus measured. If the laser beam passes through a gas plume leaking from a pipeline, the received energy will be diminished due to the absorption of the laser light in the plume. This absorption signature is used to locate the leak and assess its magnitude (Fig. 2.31).

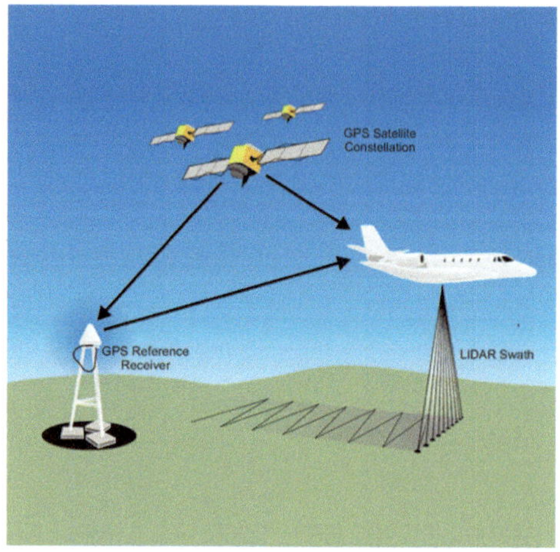

Fig. 2.31 LIDAR sensing of pipe faults

2.6.6 Ultrasonic Sensors

Ultrasonic rotating scanners are used to generate detailed three-dimensional scan of the internal pipe surface in order to detect minute deformations and cracks inside the pipe wall. Often, a laser-based excitation system is used to generate ultrasonic stress waves that can travel through the pipe (Fig. 2.32). The waveform is generally referred as guided ultrasonic wave (GUW) as the pipe boundary essentially works like a waveguide during wave transmission. Due to constructive and destructive interference, the sound intensity varies non-uniformly in the 'near-field' region. However, beyond a critical distance, the intensity becomes quite uniform, which is known as far-field region. Any flaw located in the 'far'-field region is more detectable. Piezoelectric sensors are generally used to study the nature of the wave front. In the presence of anomalies, like cracks, the waveforms get distorted which can be captured by the piezoelectric sensor. The advantage of this system is that it does not need any reference data for the prediction of anomalies. The inspection

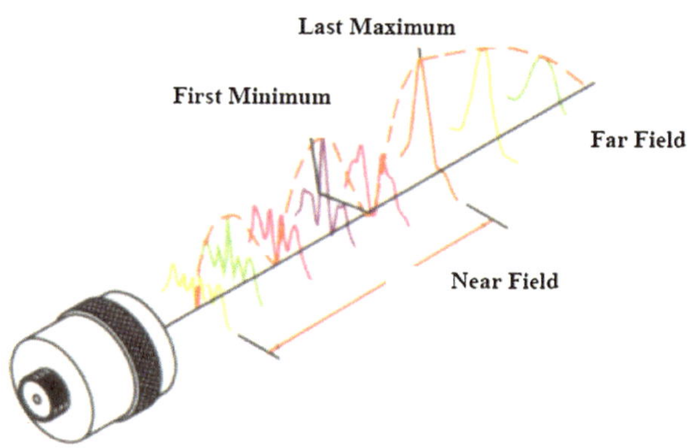

Fig. 2.32 Ultrasonic transduction system

process consists of creating a series of consecutive axial views to create a 3D image of the internal pipe wall which is later analyzed by an expert or by an automatic diagnostic system.

2.6.7 *Acoustic Leak Detection*

A leak inside a pressurized pipeline produces an acoustic signal (Fig. 2.33). This acoustic signal is created by the pressure difference inside the pipeline and the atmosphere outside the pipe. While the crawler traverses the pipeline, it continuously records this acoustic data which is analyzed to identify changes in acoustic signals. Whenever a crawler approaches, leakage intensity of the detected acoustic signal increases. The acoustic signal reaches its peak at the point at which the crawler passes the origin of the leak and then diminishes as the crawler moves away from the leak. In the following section, we will briefly compare the merits and demerits of various sensing systems.

2.7 Merits and Demerits of Various Sensing Techniques

Various sensors that we have discussed so far even though are broadly applicable for all pipe inspection robots; each sensor has its own advantages which should be considered based on application scenario and constraints. Table 2.2 provides a brief overview on relative merits and demerits.

Fig. 2.33 Acoustic leak detection system

Table 2.2 Brief performance summary of the sensors

S. No.	Sensor	Merits	Demerits
1.	Visual inspection	• Real-time monitoring • Unaffected by pipe traffic disturbance • High reliability	• Variable camera settings according to pipe size • Color of videos vary with power consumption • Requires high data storage capacity (normally 300 m/GB) • Due to high-power requirements, a power cord and video cable are always attached with the crawler
2.	Laser sensor	• Wider range of anomalies • High accuracy • High sensitivity (1 million measurements in every 4 min) • High speed of measurement	• Works only above flow lines • Power cable attached with crawler • High-power consumption • Requires additional machine vision software to generate digital pipe profile
3.	Sonar profiling	• High range • Average accuracy • Works under flow lines	• Requires supporting sensor data for accurate profiling • Power cord is attached with crawler • Signal loss occurs due to cancelation of waves reflected from different parts
4.	PVDF	• Low-power requirement • High sensitivity • Can work in rough environments • Low data output and hence less memory space requirement	• Medium–low inspection speed • More efficient in detecting positive growths • Frequent calibration is needed
5.	LIDAR	• Works best when there is existing leakage in pipeline • Highly sensitive	• Cannot be used in autonomous pipe health monitoring • High cost of operation
6.	Ultrasound sensor	• High sensitivity • Accurate modeling • Advance detection of cracks	• Large inspection time • High-power consumption • Inaccurate in the presence of certain types of anomalies like pipe joints
7.	Acoustic leak detection	• Accurate in advance leak detection • No external power is required	• Additional accessories required for position tracking • Efficient only when there is already a gas leakage

2.8 DVSS (Digital Visual Sidewall Scanning)

In the earlier sections, we have discussed the crawler robot and the sensors for carrying out automated pipe inspection. However, post-processing of the obtained data is often an equally challenging task. The experts are expected to go through extensive video data recorded for hundreds of kilometer of pipelines and detect anomalies of submillimeter size! An efficient post-processing requires an integration of efficient video-sensing (such as use of fish-eye camera, diffuse wide angle illumination) with intelligent recording, digital conversion of the data with facility for high-resolution inspection of selected critical locations. Digital visual sidewall scanning (DVSS) [19] provides an extremely reliable and convenient method for gathering visual data from within a pipe by implementing a digital image processing to deliver rich information in a format that is easy to analyze. When an area of the thumbnail is clicked, a detailed view of that region appears in the analysis pane, and a corresponding down pipe view appears alongside it. In this pane, the analyst can scroll the view in either direction and zoom in on specific regions for better inspection of the damages (Fig. 2.34).

2.9 Microwave Sensor

Microwave sensors are generally used for the inspection of thinning of the pipe wall (PWT). It detects 'invisible' damages up to 10 cm behind the pipe wall. With the help of transmitting antennas, continuous 2.45 GHz CW radar signals are sent through the pipelines. The backscattering signals obtained by the four displaced signal receiving antennas provide information regarding anomalies inside and outside of the pipe (Fig. 2.35).

Fig. 2.34 Digital visual sidewall scanning [19]

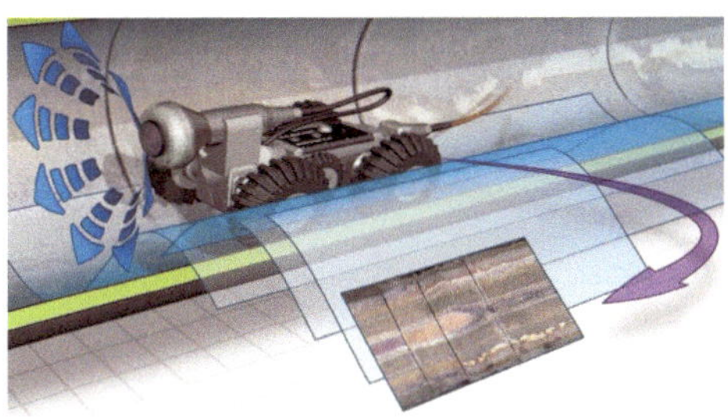

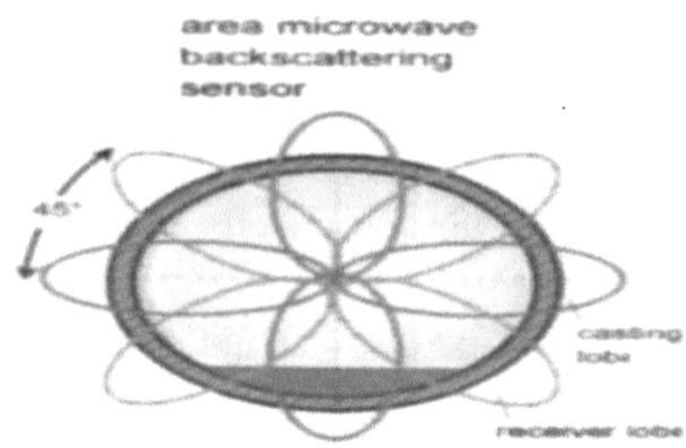

Fig. 2.35 Microwave sensing

2.10 Summary

In this chapter, we have provided an overview of pipe health monitoring systems mainly for oil and gas pipelines. We have first described the significance of such monitoring systems in view of the exponential growth of fuel pipelines and increased concern about the safety and performance of such systems. Subsequently, we have outlined the basic mechanisms corresponding to various pipe-crawling robots and focused on a few widely used models. Some of these models are later developed as multi-functional system; such that in addition to forward motion, the robots are designed to have capabilities such as adaptive size control, steering mechanism for pipe bents, and obstacle clearance system. Next, we have discussed various wheel configurations utilized with these robots and a performance comparison of such systems. Finally, we have described the sensors commonly deployed with these robots and a broad comparison of their characteristics. A few associated issues like digital signal processing systems and more advanced detecting systems based on microwave are briefly touched at the end.

Model Questions

1. What is the significance of pipeline health monitoring for oil and natural gas supply? Briefly describe the national and international scenario.
2. Write various processes/techniques that can be used for fault detection in pipelines. What is structural health monitoring? How SHM can be used for pipeline health monitoring?
3. What is in-line pipe health monitoring? What are the advantages of in-line health monitoring? Write about three types of in-line health monitoring.
4. What are traditional locomotion systems? Compare the performances between propulsion-driven PIG and belt-based PIG.
5. What are non-conventional locomotion systems? Where are they applicable? What is a Smart-ball Technology for PIG? Are Smart-balls suitable for natural gas pipelines?
6. What is an active steering system? Where are they applicable? Compare between the performances of MOGRER and FERRET robots.
7. What are multi-functional Pipe Investigating Robots? Give two examples of such PIRs and discuss their applicability.

8. What are the general challenges faced by designers for PIR locomotion? Describe two wheel assembly systems commonly used.
9. What are the various sensing mechanisms used in the pipe-crawling robots? Describe any three such sensors and indicate their advantages and disadvantages.
10. What is DVSS? How this is used for pipeline inspection?

References

1. (2012) Gas and liquid transmission pipelines. Thompson
2. http://www.nrec.ri.cmu.edu/projects/explorer/description/
3. www.robaid.com/bionics/tbcp-ii-robot-uses-gecko-tape-coated-belts-to-climb-smooth-surfaces.htm
4. http://spectrum.ieee.org/automaton/robotics/robotics-hardware/speedy-inchworm-robot-only-needs-one-motor
5. http://www.puretechltd.com/products/smartball/smartball_leak_detection.shtml
6. Okada T, Kanade T (1987) A three-wheeled self-adjusting vehicle in a pipe, FERRET-1. Int J Robot Res 6(4):60–75
7. Okada T, Sanemori T (1987) MOGRER—a vehicle study and realization for in-pipe inspection tasks. IEEE J Robot Autom RA-3(6):573–582
8. Fujiwara S, Kanehara R, Okada T, Sanemori T (1993) An articulated multi-vehicle robot for inspection and testing of pipeline interior. In: Proceedings of the 1993 IEEW/RSJ international conference on intelligent robots and systems, pp 509–516
9. Zheng Hu, Appleton Ernest (2005) Dynamic characteristics of a novel self-drive pipeline pig. IEEE Trans Rob 21(5):781–789
10. Zhang Yunwei, Yan Guozheng (2007) In-pipe inspection robot with active pipe-diameter adaptability and automatic tractive force adjusting. Mech Mach Theory 42(12):1618–1631
11. Kim YJ, Yoon KH, Park YW (2009) Development of the inpipe robot for various sizes. In: IEEE/ASME international conference on advanced intelligent mechatronics, pp 1745–1749
12. Roh S, Kim DW, Lee JS, Moon H, Choi HR (2009) In-pipe robot based on selective drive mechanism. Int J Control Autom Syst 7(1):105–112
13. Yu-xia Z, Guang-ping H, De-wen G (2009) Micro in-pipe robot mechanical structure design of shape memory alloy driving. In: Proceedings of the 2009 IEEE international conference on robotics and biomimetics, 19–23 Dec 2009, Guilin, China
14. Choi HR, Roh S (2007) In-pipe robot with active steering capability for moving inside of pipelines. In: Habib MK (ed) Bioinspiration and robotics: walking and climbing robots. ISBN 978–3-902613-15-8, p 544, I-Tech, Vienna, Austria, EU, Sept 2007
15. Matos J, Tosunoglu S (2011) Peristaltic crawler for piper-line unplugging. ASME Early Career Tech J. In: ASME early career technical conference, ASME ECTC 4–5 Nov 2011, Atlanta, Georgia USA
16. Choi HR, Ryew SM (2002) Robotic system with active steering capability for internal inspection of urban gas pipelines. J Mechatron 12(5):713–736
17. Liu Zheng, Kleiner Yehuda (2013) State of the art review of inspection technologies for condition assessment of water pipes. Measurement 46:1–15
18. Agarwal V, Ogai H, Nishijima K, Bhattacharya B (2011) Inspection of pipe inner surface using advanced pipe crawler robot with PVDF sensor based rotating probe. Sens Transducers 4(127):45–56
19. http://www.pipelineandgasjournal.com/digital-visual-sidewall-scanning-streamlining-collection-and-analysis-inspection-data

Chapter 3
Basic Mechanical Design of Passively Transported Pipe Inspection Robots

3.1 Introduction

In the previous chapter, brief descriptions of various types of mechanisms related to pipe crawling robots are provided [1–7]. However, for the beginner to carry out the mechanical design of pipe-crawling robots, one needs to consider an elementary problem of moving a pipe crawling robot in a horizontal gas pipe network where the pressure of the gas can provide necessary kinetic energy to the robot locomotion. The same problem can also be extended to other fluid transportation systems like water pipes. In this chapter, we will explain the preliminary design principles which are necessary for a robot to maneuver inside a pipe. This preliminary design involves kinematic analysis, mass estimation, and suspension design. Finally, we will address two advanced topics related to mechanism design and dynamic modeling of the system.

3.2 Kinematic Analysis of Robot Module

A pipe crawling robot often has to pass through bends and T-joints of pipelines. Hence, while designing a robot, its dimensions should be considered in a way that the robot can easily pass through various bends in the pipeline. In this section, we will discuss the design consideration which ensures easy maneuver of the robot through the bends. In the adjacent Fig. 3.1, various design parameters are explained which influences the geometry of the robot. Also, the figure shows the position of the robot while the extreme dimension of allowable length and width is considered, i.e., if we increase the length of the robot, then we have to decrease the width and vice versa.

H. Ogai and B. Bhattacharya, *Pipe Inspection Robots for Structural Health and Condition Monitoring*, Intelligent Systems, Control and Automation: Science and Engineering 89, https://doi.org/10.1007/978-81-322-3751-8_3

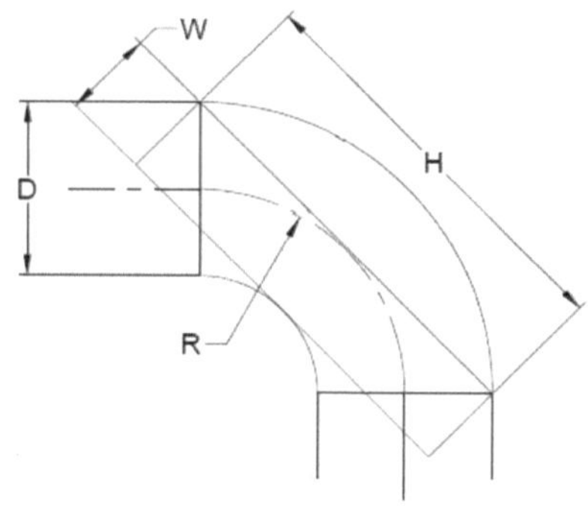

Fig. 3.1 Schematic of the pipe elbow

Following are the parameters used in finalizing the geometry of the robot:

H Length of robot
W Width of robot
R Radius of bend
D Inner diameter of the pipe

Based on the extreme position, two cases may arise. In the following sections, designing of robot in these two conditions is explained.

Case 1 Length of robot is more than the extreme length (i.e., AD in Fig. 3.2). The adjacent figure shows the condition when the length of the robot is more than the extreme length. For a robot to pass through the 90° bend, it will be necessary to reduce the width of the robot. Design of robot for such condition is explained below.

From the Fig. 3.2, we know that:

$$\angle BOQ' = 45^\circ$$

Also, from geometry we can find,

$$\begin{aligned} AB = NQ' &= [OB\cos 45^\circ - ON] \\ &= \left[\left(R + {}^{D}\!/_{2}\right)\sin 45 - \left(R - {}^{D}\!/_{2}\right)\right] \end{aligned}$$

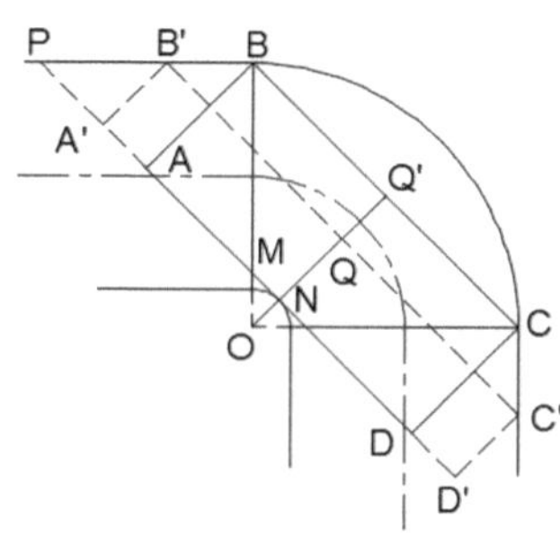

Fig. 3.2 Diagram illustrating Case 1

The parameters R and D are already explained in Fig. 3.1. Consider $A'B'$ to be the new width of the robot. For the robot to pass through the bend, the length of the robot should be smaller than $A'D'$.

Now, the constraint on $A'D'$ can also be written as,

$$A'D' \leq 2 \times A'N$$

$$\begin{aligned} A'N &= MN + PM - A'P \\ &= MN + 2AB - A'B' \\ &= \left(R + D/2\right) + 2 \times \left(R + D/2\right)\cos 45 - R - D/2 - A'B' \end{aligned}$$

$$A'D' \leq \left[R - D/2 + 2\left\{\left(R + D/2\right)\cos 45 - \left(R - D/2\right)\right\} - A'B'\right] \times 2$$

The above condition may be further simplified and written as follows:

$$A'D' \leq \left[2R\left(\sqrt{2} - 1\right) + D\left(\sqrt{2} + 1\right) - 4A'B'\right]$$

Case 2 Width of the robot is more than the extreme Width (i.e., AB).
The adjacent Fig. 3.3 shows the case, when width of the robot is more than the extreme width, the length of the robot should be restricted below the extreme length. This will ensure that robot can easily pass through the bends/elbows.

Now, $A'D'$ can be written as,

$$A'D' \leq 2 \times B'Q$$

$$\begin{aligned} A'N &= \sqrt{B'O^2 - OQ^2} \\ &= \sqrt{\left(R + D/2\right)^2 - \left(R - D/2 + A'B'\right)^2} \end{aligned}$$

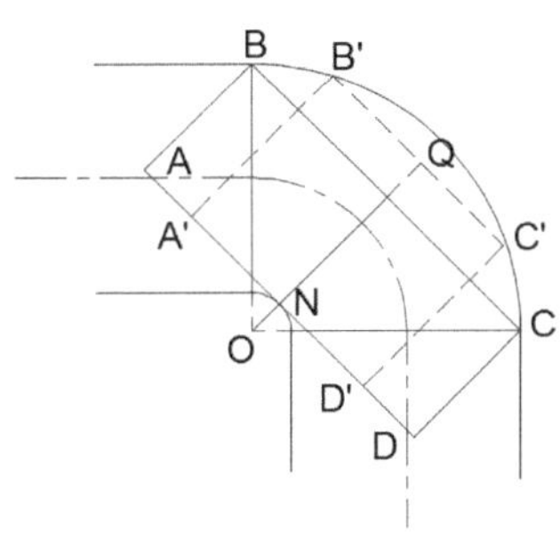

Fig. 3.3 Diagram illustrating Case 2

$$A'D' \leq \sqrt{\left(R + {}^{D}\!/_{2}\right)^{2} - \left(R - {}^{D}\!/_{2} + A'B'\right)^{2}} \times 2$$

To summarize,
If $A'D' = H'$ and $A'B' = W'$, then for

$$0 < W' \leq \left[\left(R + {}^{D}\!/_{2}\right)\cos 45 - \left(R - {}^{D}\!/_{2}\right)\right]$$

$$H' \leq \left[R - {}^{D}\!/_{2} + 2\left\{\left(R + {}^{D}\!/_{2}\right)\cos 45 - \left(R - {}^{D}\!/_{2}\right)\right\} - W'\right] \times 2$$

And for,

$$\left[\left(R + {}^{D}\!/_{2}\right)\sin 45 - \left(R - {}^{D}\!/_{2}\right)\right] < W' \leq D$$

$$H' \leq 2 \times \sqrt{\left(R + {}^{D}\!/_{2}\right)^{2} - \left(R - {}^{D}\!/_{2} + W'\right)^{2}}$$

We will further elucidate the application of the constraint for estimating the size of the robot with the help of the following problem.

Problem 3.1 A pipe-crawler robot is designed whose width is 180 mm and length 300 mm. Check whether the robot can pass through a 90° pipe bend of internal diameter 200 mm and radius of curvature 200 mm.

Solution
Given,

$$\begin{aligned} \text{Length of robot}, H &= 400\,mm \\ \text{Width of robot}, W &= 180\,mm \\ \text{Radius of bend}, R &= 200\,mm \\ \text{I.D. of pipe}, D &= 200\,mm \end{aligned}$$

For the above dimension of pipe bend, maximum allowable width of the robot should be, W'

$$\begin{aligned} &= \left[\left(R + {}^{D}\!/_{2}\right)\cos 45 - \left(R - {}^{D}\!/_{2}\right)\right]. \\ &\cong 112\text{ mm} \end{aligned}$$

Since width of the robot is more than the maximum allowable width, length of the robot should be as per **Case 2**. i.e., length of the robot should be less than or equal to:

$$H' = 2 \times \sqrt{\left(R + {}^{D}\!/_{2}\right)^2 - \left(R - {}^{D}\!/_{2} + W'\right)^2} \cong \mathbf{560\ mm.}$$

Since actual length (400 mm) of robot is less than maximum allowable length (560 mm), the robot will easily pass through 90° pipe bend of internal diameter 200 mm and radius of curvature 200 mm.

It may be noted that often due to the functional requirement you may not limit the length, in such cases you have to redesign the robot as a modular robot consisting more than one modules and joined using flexible joint system.

3.3 Estimation of Mass of the Robot

For designing a robot, mass estimation is the next important factor. The mass of the robot needs to be as per with the drag force. If the mass of the robot exceeds allowable mass, then the drag forces generated by the fluid flowing will not be sufficient enough to move the robot forward. In this section, stepwise calculation procedures required for determining allowable mass of robot are explained. To explain the procedure, a three-wheel-based robot is considered. Driving force of the robot is provided by the fluid flowing inside the pipe. This calculation steps can be extended for determining weight of any numbers of wheel-based system.

The major driving forces to propel the module forward are the drag forces. Its forward motion is hindered by Resistive Rolling Friction. In order to calculate the maximum payload that the module can carry and move with a constant velocity, the force balance between drag force and the Resistive Rolling Frictional force has been balanced.

First, we calculate the drag forces experienced by the crawler by the relative motion of the fluid flowing past it.

3.3.1 Analysis of Drag Forces on the Crawler

Figure 3.4 shows the schematic of the crawler robot and various external forces working on it.

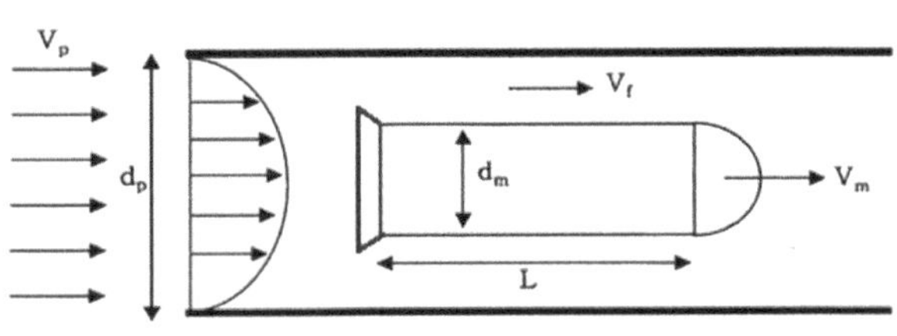

Fig. 3.4 Schematic of the crawler inside a pipe with fluid flow

The relevant parameters that affect the drag force induced on the robot are indicated in the figure. Following is the description of those parameters involved:

d_p	Pipe diameter
d_m	Module diameter
d_b	Module end diameter (back-plate projection)
d_h	Hydraulic diameter ($=d_m/2$)
V_f	Pipe flow velocity
V_p	Average velocity between module and pipe
L	Length of the module
V_m	Velocity of the module
C_d	Drag coefficient
ρ_{fluid}	Density of fluid
u_∞	Relative velocity
A	Projected area
M	Mass of the robot module

For the ease of calculation, we have considered the robot body to be cylindrical.

Following previous assumptions, a simple drag analysis is performed. This includes a pressure drag caused by the pressure difference between the front and rear part of the robot and a friction drag that results from the boundary layer formed along the surface of the robot. Accordingly, total drag force may be defined as follows:

$$\text{Drag force} = D_{pressure} + D_{friction}. \tag{3.1}$$

The pressure drag ($D_{pressure}$) can be calculated from the following equation, with the appropriate value of drag coefficient (C_d) chosen based on the shape of the robot.

$$D_{pressure} = \frac{1}{2} C_d \rho_{fluid} u_{\infty_1}^2 * A_{pressure} \tag{3.2}$$

where

$$\begin{aligned} u_{\infty_1} &= V_f - V_m \\ u_{\infty_2} &= \frac{V_f \pi (d_p)^2}{\pi\left((d_p)^2 - (d_m)^2\right)} - V_m \\ A_{pressure} &= \pi \left(\frac{d_b}{2}\right)^2 \end{aligned}$$

Next, we will calculate the skin friction drag (D_{friction}) as follows:
Using Darcy–Weisbach equation:

$$D_{\text{friction}} = f_{\text{D}} * 0.5 * u_{\infty_2}^2 * \rho * \left(\frac{L}{d_{\text{h}}}\right) * A_{\text{friction}} \tag{3.3}$$

$$f_{\text{D}} \cong f_{\text{moody}}(Re, k, d_{\text{h}})$$

where f_{D} = skin friction coefficient, Re = Reynold's number, k = equivalent roughness, f_{moody} = friction factor from Moody's chart

$$A_{\text{friction}} = \pi * d_{\text{m}} * L.$$

3.3.2 *Quasi-static Analysis of the Robot*

Figure 3.5 shows the free body diagram (FBD) of the robot, where N_1 and N_2 are the reaction caused by spring force and weight of the robot, respectively.
Spring force on each spring can be written as follows:

$$N_1 = Kx \tag{3.4}$$

where K is the spring stiffness and x is the deformation.
From static equilibrium:

$$N_1 + Mg = 2(N_1 + N_2)\cos 60^\circ \tag{3.5}$$

Now,

$$N = 3N_1 + 2N_2. \tag{3.6}$$

Fig. 3.5 FBD of the robot

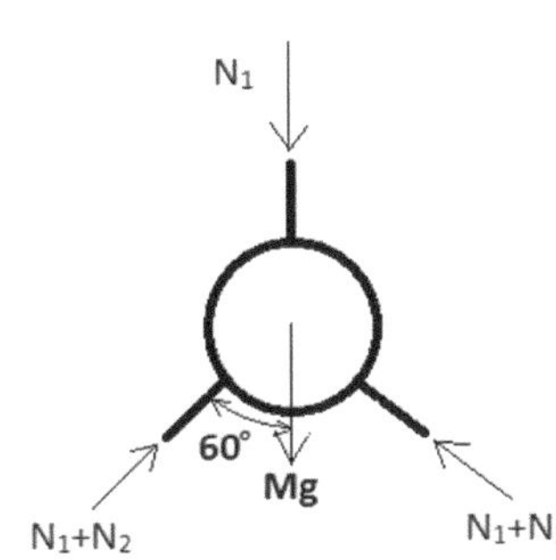

3.3.3 Analysis of Rolling Friction Resistance

Thus, the equation of motion for the robot is given by:

$$M \times a = F_{\text{drag}} - F_{\text{resistive}}$$

For, the crawler to move with constant velocity, $a = 0$

$$F_{\text{drag}} = F_{\text{resistive}}$$

The rolling friction resistance from the wheels may be expressed as follows:

$$F_{\text{resistive}} = C_{\text{rr}} N,$$

where

$F_{\text{resisitive}}$	Rolling resistance force
C_{rr}	Coefficient of rolling friction (CRF)
N	Force perpendicular to the surface on which the wheel is rolling

From the above discussion, we can form the governing equation as,

$$D_{\text{pressure}} + D_{\text{friction}} = C_{\text{rr}}(3N_1 + 2N_2) \tag{3.7}$$

Using Eqs. (3.1)–(3.7), allowable weight of the robot to traverse passively with the help of kinetic energy of the fluid can be determined.

Problem 3.2 Estimate the maximum permissible mass of a robot given the following system parameters such that the robot will be able to move with the flow with uniform velocity. Consider the robot is supported by three-spring-loaded (stiffness of each spring, k = 10 N/mm) wheel assembly placed 120° apart, and the maximum deformation in the spring is 2 mm.

Given parameters

Pipe diameter, d_{p} = 0.2032 m
Module end diameter, d_{b} = 0.180 m
Average module diameter, d_{m} = 0.1 m
Module length, L_{b} = 0.32 m
Pipe flow velocity, V_{f} = 0.8 m/s
Flow density, ρ = 27.412 kg/m^3
Fluid viscosity, $\mu = 16 \times 10^{-6}$ N.s/m^2
Drag coefficient, C_{d} = 0.05
Module velocity, V_{m} = 0.3 m/s

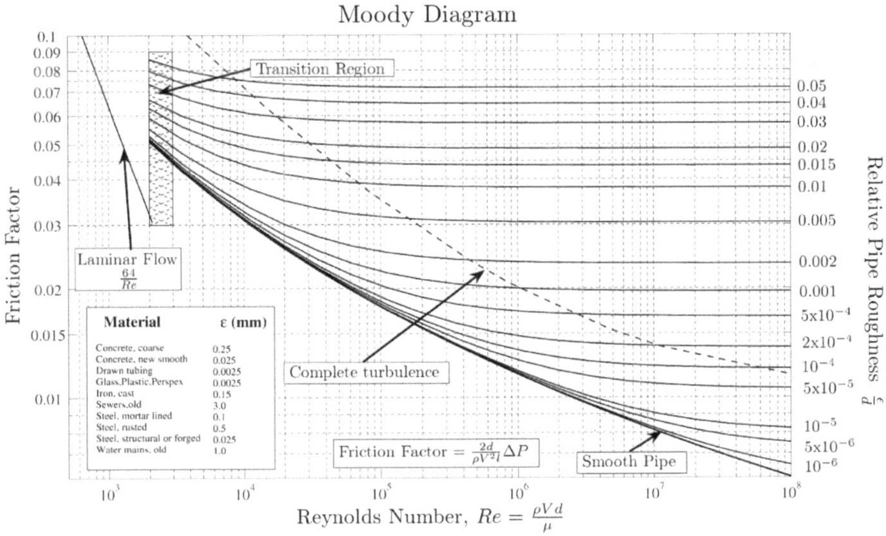

Fig. 3.6 Moody's diagram

Solution

For the hydraulic diameter $d_m = 0.05$ m, the Reynold's number is calculated to be $1.31 * 10^{-5}$.

Correspondingly from the Moody's chart (Fig. 3.6), $f_m = 0.015$.

$$D_{\text{pressure}} = 0.3312 * u_{\infty_1}^2$$

$$D_{\text{friction}} = 0.0914 * u_{\infty_2}^2$$

$$u_{\infty_1}^2 = 0.8 - V_m$$

$$u_{\infty_2}^2 = 1.066 - V_m$$

Assuming rolling friction coefficient = 0.0025 (between rubber wheel and steel pipe),

Since $D_{\text{friction}} \ll 1$, this can be neglected.

For zero acceleration, the resistive force due to rolling friction will be equal to the net drag force.

We get,

$$F_{\text{resistive}} = 9.8 C_{\text{rr}} M.$$

For zero acceleration, the resistive force due to rolling friction will be equal to the net drag force. Hence,

$$9.8C_{rr}M = D_{\text{pressure}} + D_{\text{friction}}$$

using the expressions for D_{pressure} and D_{friction}, derived earlier, we get:

$$M = 13.52u_{\infty_1}^2 + 1.865u_{\infty_2}^2$$

$$M = 13.52(0.8 - V_m)^2 + 1.865(1.066 - V_m)^2$$

You can easily find out that for $V_m = 0.3$ m/s, $M = 4.476$ kg, for $V_m = 0$, $M = 10.775$ kg.

Hence, the maximum allowable mass = **4.476 kg**.

3.4 Stress Analysis Between the Two Modules of a Robot

Whenever the total mass of the robot exceeds the allowable mass, we need to divide it into multiple modules. In such cases, the joint between the two modules is to be analyzed against two tip loads.

For example, let us consider a solid cylindrical joint of length '*L*' having two equal weight modules at its two ends as shown in Fig. 3.7. Assuming 'δ' as deflection, we can find the same for a 60° bend with radius of curvature '*r*' as follows:

$$\Delta = r - r\cos 30°; \quad r(\text{radius}) = \frac{6L}{2\pi}$$

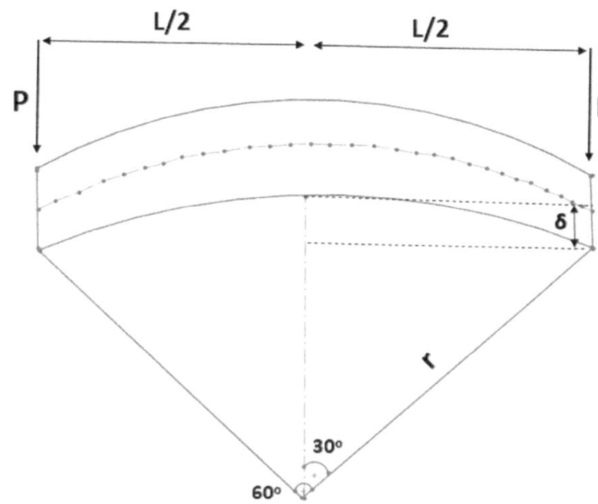

Fig. 3.7 Deflection at the joint

We can then obtain the deflection as follows:

$$\delta = \left(\frac{6L}{2\pi}\right) * (1 - \left(\frac{\sqrt{3}}{2}\right) = 0.13L.$$

We also know from the force-deflection relationship in a cantilever beam that

$$\delta = \frac{P\left(\frac{L}{2}\right)^3}{3\mathrm{EI}} = 0.13L,$$

Hence,

$$P = \left(\frac{3 * 0.13 * 8 * \mathrm{EI}}{L^2}\right)$$

where EI is the flexural rigidity of the link.

We can, subsequently, calculate the maximum stress as follows:

$$\sigma_{max} = \left(\frac{PL}{2Z}\right) = \left(0.39 * \frac{8\mathrm{EI}}{2LZ}\right)$$

where Z is the section modulus.

Since we have two cantilevers as the joining elements between the two modules, therefore, in the worst case scenario, the stress will be doubled as follows:

$$\sigma_{\mathrm{critical}} = 3.12\mathrm{EI}/(LZ) \tag{3.8}$$

Also, due to the thrust on the module, the compressive stress

$$\sigma_c = \frac{F}{\pi(r_0^2 - r_i^2)} \tag{3.9}$$

where 'F' is the force on the end plate:

$$F = \frac{500 * \pi(0.18)^2}{4} = 12.7\ \mathrm{N}$$

Hence, the compressive stress on the bottom surface of the link:

$$\sigma_{\mathrm{net_{comp}}} = \sigma_{\mathrm{crit}} + \sigma_{\mathrm{c}}$$

$$\sigma_{\mathrm{net_{comp}}} = 3.12\frac{\mathrm{EI}}{LZ} + F/\pi(r_0^2 - r_i^2) \tag{3.10}$$

Thus, the corresponding design condition may be written as follows:

$$\sigma_{\mathrm{net}_{\mathrm{comp}}} \leq S_{y_{\mathrm{comp}}} \tag{3.11}$$

While for the top surface of the link, the tensile stress is:

$$\sigma_{\mathrm{net}} = \sigma_{\mathrm{crit}} - \sigma_{\mathrm{c}}$$

$$\sigma_{\mathrm{net}} = 3.12\frac{\mathrm{EI}}{LZ} - F/\pi(r_0^2 - r_i^2) \tag{3.12}$$

Thus, against the tensile stress, the corresponding design condition may be written as follows:

$$\sigma_{\mathrm{net}} \leq S_{y_{\mathrm{tension}}}. \tag{3.13}$$

3.5 Suspension Design for the Wheels

The pipe crawling robots are generally integrated with various types of defect sensors. In order to have a noise-free data recording, smooth travel of the robot is absolutely essential. This, in turn, requires a good suspension design for the robot.

Since each module is supported by three-spring-loaded system, each running over a section of the same pipe surface; we can consider an equivalent spring-dashpot model such that the model may be considered to be a single degree of freedom system moving over a rough surface (as shown in Fig. 3.8).

Accordingly, the equivalent stiffness and damping may be written in the frequency domain as follows:

$$K_{\mathrm{eqv}} = \sum_{i=1}^{3}[k_i] \tag{3.14}$$

$$c_{\mathrm{eqv}} = \sum_{i=1}^{3}[sc_i] \tag{3.15}$$

where k_i is the stiffness of individual spring and c_i is the damping of each individual damper. Here, we have considered a Kelvin–Voigt representation of stiffness and damping in parallel connection. Now the final governing equation in the steady state may be written as follows:

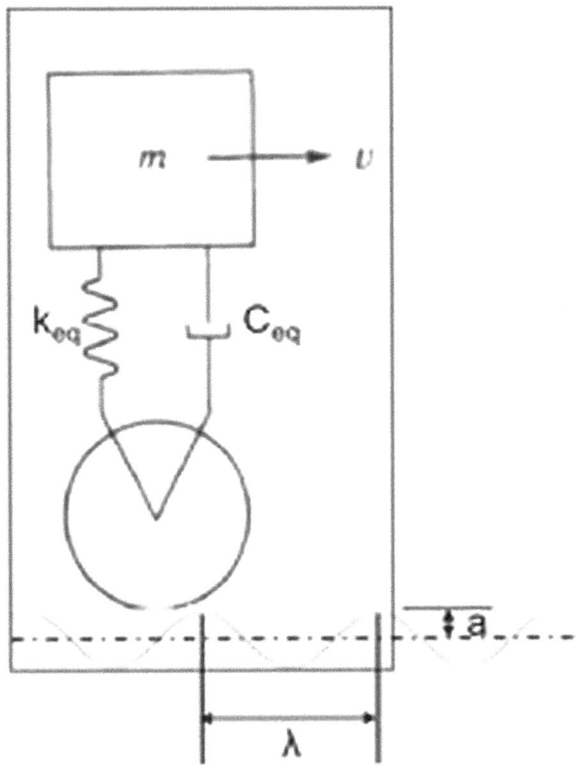

Fig. 3.8 Equivalent suspension model of the robot

$$\left[s^2 m + s c_{\text{eqv}} + k_{\text{eqv}}\right] X(s) = \frac{a\omega}{(s^2 + \omega^2)} \tag{3.16}$$

Here, $\omega = 2 * \pi * v/\lambda$ such that, v = the velocity of the module, a = amplitude of surface roughness, and λ = the wavelength of the surface roughness. Equation (3.16) is a classic problem of harmonic excitation which may be solved iteratively for suitable values of the stiffness and damping coefficient until the displacement amplitude falls below the maximum tolerable displacement of the system.

3.6 Traction Force for Active Drive Mechanism

In the previous section, passive drive mechanism has been explained. In this section, focus will be on active drive mechanism. In the active drive mechanism, motorized robot wheels are used to drive the robot forward. Since wheels are externally powered, there should be enough traction force between wheels of robot and the pipe wall; otherwise, the wheels will slip and efficiency of the robot will

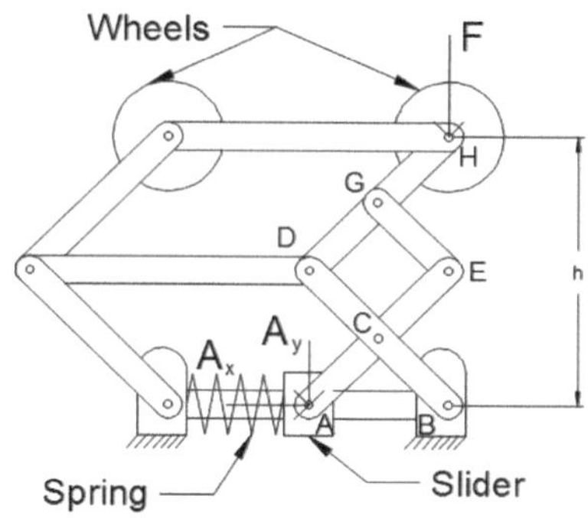

Fig. 3.9 Pantograph wheel mechanism

decrease. In Fig. 3.9, an example of pantograph wheel mechanism is being shown. It can be seen that the traction force between wall and wheel is provided by a spring.

It is important to design this mechanism with appropriate spring stiffness, since if the spring is less stiff than required, then the wheel will slip; if the stiffness is more, then higher power will be required to drive the wheel. Hence, in this section, determining spring stiffness for the pantograph mechanism will be discussed.

Let the normal force on the wheel be '*F*,' the traction force is proportional to this normal force. Also, let '*K*' be the stiffness of the spring.

Also, link length, $DH = BD = AE = l$, and '*G*' and '*C*' are the midpoints of link *DH* and *DB*, respectively. Hence, from the geometry we can say that,

$$\angle HDE = \angle EAB = \angle DBA = \angle GED = \theta \tag{3.17}$$

Let the initial length of *AB* be '*x*,' then

$$x = l\cos\theta$$

$$\therefore \; \mathrm{d}x = -l\sin\theta\mathrm{d}\theta$$

Also,

$$\frac{h}{2} = l\sin\theta$$

$$\therefore \; \mathrm{d}h = 2l\cos\theta\mathrm{d}\theta$$

A_x and A_y are the horizontal and vertical components of force on point 'A.' Since point A is restricted to move vertically, the force 'F' will depend only on the horizontal movement of spring, and hence, A_y can be ignored.

From the principle of virtual work, we can say that,

Total work done by the forces to move some virtual displacement is zero.

Hence,

$$\text{Total work Done} = F \times \partial h + A_x \times \partial x$$

$$0 = F \times 2l \cos\theta - A_x \times l \sin\theta$$

$$A_x = 2F/\tan\theta \tag{3.18}$$

Also,

$$A_x = K \times \delta x$$

$$\therefore\ K \times \delta x = 2F/\tan\theta \tag{3.19}$$

where 'δx' is the deformation of spring due to change in 'h.'

Problem 3.3 For the pantograph type wheel assembly, determine the normal force on the wheel if the link length is 50 mm, h = 40 mm, spring constant, K = 100 N/mm, and initial deformation of the spring due to changing h is, $\delta x = 10$ mm

Solution
Since l = 50 mm, h = 40 mm

$$\therefore\ \tan\theta = \frac{40}{(50^2 - 40^2)^{.5}} = \frac{4}{3}$$

We know that,

$$\therefore\ K \times \delta x = 2F/\tan\theta$$

$$F = 666.67\ \text{N (Approx)}.$$

Summary

In this chapter, we have carried out the basic mechanical design of pipe crawling robots. We have initially considered the kinematics of the motion, corresponding to the critical motion of the robot across a 90° bend in the pipe and obtained the conditions on length and width such that the robot will not get stuck inside the pipe. We have then shown, how to estimate the maximum allowable mass of the robot based on the drag computation. If the payload requires additional mass for the robot, then we generally go for modular design. In case of modular robots, the joints are the weakest link of the system, and hence, we have shown the design consideration for the maximum stress analysis at the joint. Next, we have discussed the suspension design considering the dynamic model of the system. In this design, we have considered equivalent stiffness and damping contribution from the three wheels in the form of a Kelvin–Voigt model. Finally, we have discussed an active mechanism for powered crawler robot and have shown the process of computation of spring stiffness by using virtual work principle.

Model Questions

1. Find out the governing equation of the robot system using energy formulation and Lagrange's equation
2. What are the critical design parameters for the design of pipe-crawler robot against force-balanced system?
3. What are the different ways by which we can reduce the mass of a pipe-crawling robot?
4. Solve Eq. (3.16) for a pipe crawling robot to obtain the displacement and acceleration amplitudes.
5. Consider any advanced mechanism for traction of the pipe crawling robot and find out the link forces.

References

1. Okada T, Kanade T (1987) A three-wheeled self-adjusting vehicle in a pipe, FERRET-1. Int J Rob Res 6(4):60–75
2. Okada T, Sanemori T (1987) MOGRER-A vehicle study and realization for in-pipe inspection tasks. IEEE J Robot Autom RA-3(6):573–582
3. Fujiwara S, Kanehara R, Okada T, Sanemori T (1993) An articulated multi-vehicle robot for inspection and testing of pipeline interior. In: Proceedings of the 1993 IEEW/RSJ international conference on intelligent robots and systems, pp 509–516
4. Hu Z, Appleton E (2005) Dynamic characteristics of a novel self-drive pipeline pig. IEEE Trans Rob 21(5):781–789
5. Zhang Y, Yan G (2007) In-pipe inspection robot with active pipe-diameter adaptability and automatic tractive force adjusting. Mech Mach Theory 42(12):1618–1631
6. Kim YJ, Yoon KH, Park YW (2009) Development of the in-pipe robot for various sizes. In: IEEE/ASME international conference on advanced intelligent mechatronics, pp 1745–1749
7. Roh S, Kim DW, Lee JS, Moon H, Choi HR (2009) In-pipe robot based on selective drive mechanism. Int J Contr Autom Syst 7(1):105–112

Chapter 4
Experiments of Wireless Transfer Technology for Communication

4.1 Introduction

Pipelines related to various industrial applications usually degrade within a few years due to erosion and corrosions, such as steel pipes in factories, PVC/steel pipes used for water and gas supply. Similar deteriorations also occur for ceramic pipes, concrete pipes, and other metallic pipes. In order to inspect and repair these pipes, flexible automatic inspection robot is needed. In addition, for controlling the inspection robot and transmitting the collected data to terminal station which is located at the pipe inlet, wireless communication system is essential [1]. In this chapter, we will discuss about our experiments with wireless transfer technology for real-time maintenance of pipelines and present a performance study of this system for different pipe health monitoring conditions.

In 2005, radio wave transmission properties in steel pipes were measured at Ogai laboratory [2–4]. Wireless LAN with orthogonal frequency division multiplexing (OFDM) of frequencies 2.4 and 5.2 GHz was tested, respectively. It was observed that for the diameter of measured steel pipe between 20 and 40 cm, and length within 100 m, more than 10 Mbps information could be transmitted, the loss was minimal, and the pipes can work as waveguide. Subsequently, in 2016, similar properties in steel pipes were measured for Wi-Fi with 2.4 GHz by using two types of access points (IITK campus Wi-Fi and Airtel Wi-Fi). The diameter of measured steel pipe in this case was 20.3 cm, and the distance was 25 m. The transmission loss was about 10 dB in 25 m at IIT, Kanpur.

Further, transmission property of ceramic pipe with a diameter of 20 cm was also tested in Japan. It was found that wireless transmission was greatly affected by pollutant while compared with a wireless transmission test for similar pipe inside the laboratory. In the presence of smaller pollutant in ceramic pipe, antenna gain and transmission output would increase significantly. This will be discussed in Sects. 4.2.2 and 4.3.

H. Ogai and B. Bhattacharya, *Pipe Inspection Robots for Structural Health and Condition Monitoring*, Intelligent Systems, Control and Automation: Science and Engineering 89, https://doi.org/10.1007/978-81-322-3751-8_4

In 2007, other measurements in 20-m ceramic pipes with different diameters were carried out at 5.2 GHz frequency. From the measurements, transmission loss was found to be 86 dB for 25 cm diameter and 65 dB for 30 cm diameter, respectively. Interestingly, more loss was observed in the 25-cm-diameter ceramic pipe. This may happen in smaller pipes as there are more reflections from the wall during wave propagation, and consequently, more absorption takes place.

In this chapter, experiments related to the wireless communication system are explained at length. It is shown that the wireless communication system can steadily transmit image data at high speed (10 Mbps). On the basis of wireless communication system, a new pipe inspection system for drain pipe is designed. With this system, remote control of robot movement is realized, defects or troubles inside the pipe could be inspected, and also image information inside the pipe can be transmitted in real time.

4.2 Pipe Inspection System

4.2.1 Transmission Test in Steel Pipe

For the first experiment, a 10-m steel pipe with a diameter of 30 cm is chosen as shown in Fig. 4.1. A signal generator is placed in the pipe inlet, and an antenna and a spectrum analyzer are located at the outlet. The transmission properties were then tested. Figures 4.2 and 4.3 show the transmission loss at different antenna positions for 2.4 and 5.2 GHz frequency bands, respectively.

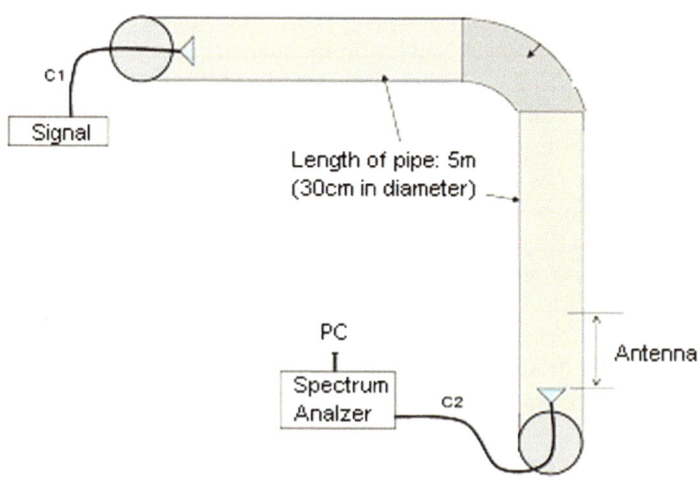

Fig. 4.1 Wireless communication property measurement in steel pipe

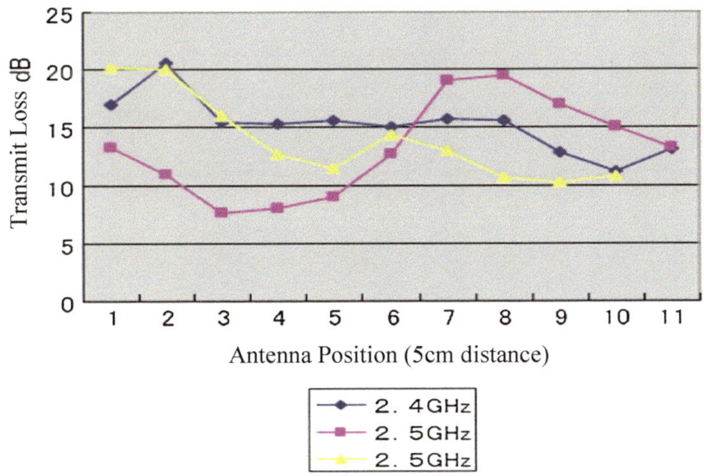

Fig. 4.2 Transmission loss measured in steel pipe (2.4 GHz)

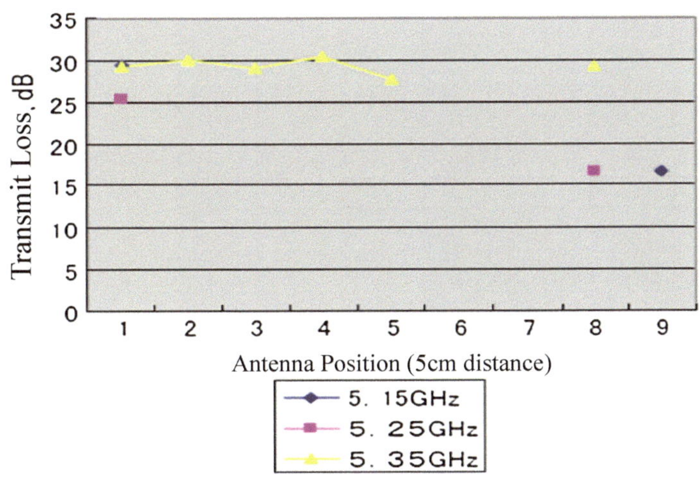

Fig. 4.3 Transmission loss measured in steel pipe (5.2 GHz)

A 25-m steel pipe with a diameter of 20.3 cm is shown in Fig. 4.4. Wi-Fi with 2.4 GHz was tested by using two types of access points. One was by using IITK campus Wi-Fi, and the other was an Airtel Wi-Fi system. The measured radio field intensity by using campus and Airtel Wi-Fi are shown in Fig. 4.5a and b, respectively.

Fig. 4.4 A 25-m-long steel pipe for wireless communication test at IIT, Kanpur

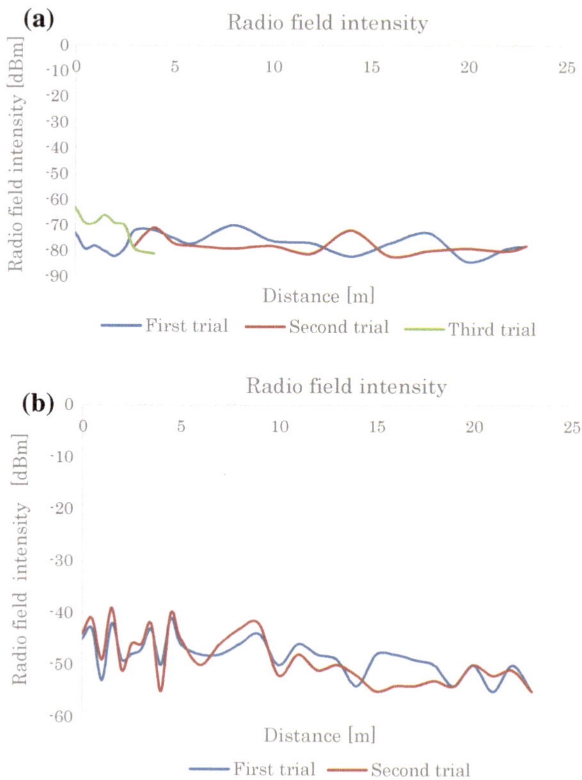

Fig. 4.5 **a** Radio field intensity using campus Wi-Fi (2.4 GHz). **b** Radio field intensity using Airtel Wi-Fi (2.4 GHz)

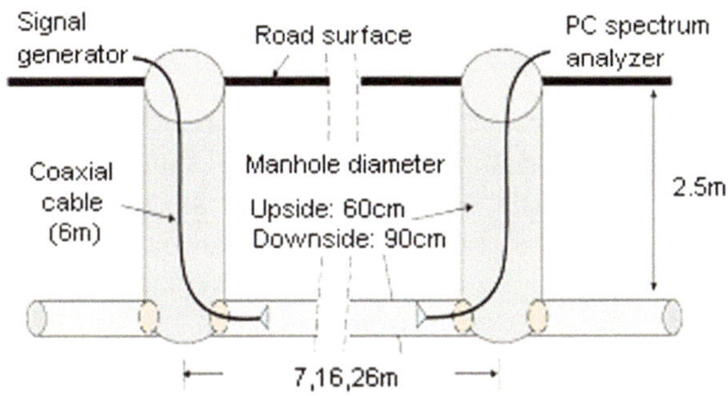

Fig. 4.6 Measurement of properties of wireless communication in a ceramic pipe

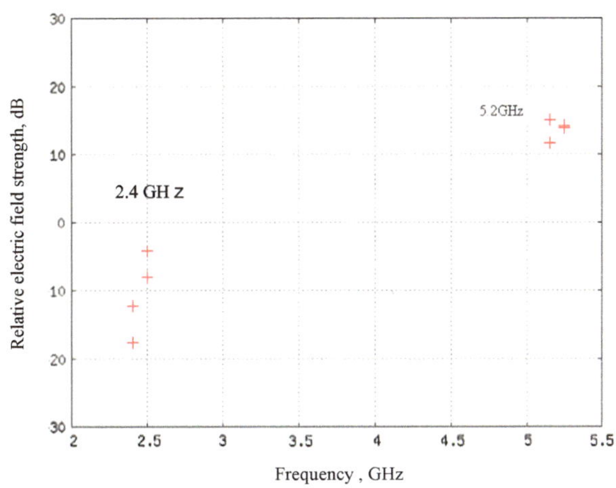

Fig. 4.7 Results of measuring wireless communication in a 7-m-long ceramic pipe

4.2.2 Transmission Test in Ceramic Pipe

A schematic diagram of a ceramic pipe with a diameter of 20 cm and a length of 7 m is illustrated in Fig. 4.6. The transmission loss was 84 ± 2.5 dB in the 2.4 GHz frequency band and 52 ± 1.5 dB in the 5.2 GHz frequency band, respectively. Figure 4.7 shows the relative electric field strength between the ceramic pipe and the free space. It is clear that the 5.2 GHz band shows better performance in a ceramic pipe.

4.3 Robot Test and Transmission Loss Measurement in Actual Ceramic Drainage Pipe in Japan

Measurements of the transmission loss in an actual ceramic drainage pipe were carried out at Yahata in the city of Kitakyushu in 2007.

4.3.1 Robot Test

The wireless robot could move inside the pipe, and it transmitted images of the interior in real time. Figure 4.8 [2, 3] shows the pipe inspection robot used for this purpose. This robot consists of a wireless system with 5.2 GHz frequency, a lithium-ion battery system, a web camera, a moving motor, and a control system. Figure 4.9 shows an image of the interior of the pipe obtained using this wireless robot.

We test the radio communication system in an actual drain pipe at Momozono in Kitakyushu city. The wireless robot could run in the pipe with 20 m length and transmit the inside image data in real time.

Fig. 4.8 Robot pipe inspection robot system

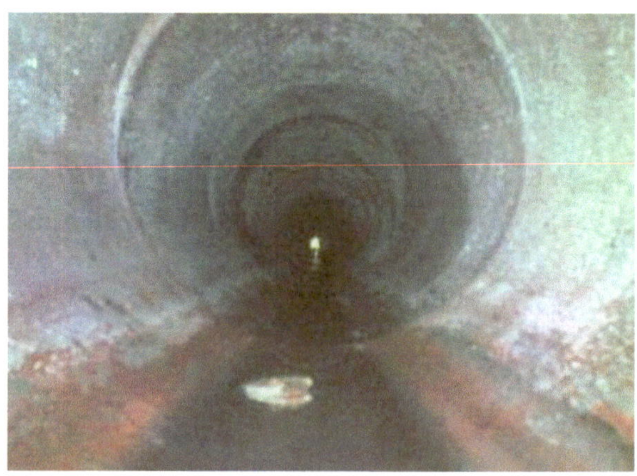

Fig. 4.9 Image inside pipe

4.3.2 Transmission Loss Measurement

The measurements of transmission loss were carried out under the following conditions:

(1) the ceramic pipes were well cleaned.
(2) the length of the test section was 19 m.

Two types of ceramic pipes with different diameters of 25 cm and 30 cm were measured.

Figure 4.10 is a schematic diagram of the measurement system used for obtaining the transmission loss.

4.3.3 Measurement Results [4]

Figure 4.11 shows the electric power output. The transmission loss and their approximate equations for different pipes are shown in Fig. 4.12. Transmission loss between antenna points, which is measured in the laboratory, is illustrated in Fig. 4.13. It was found that the transmission loss has a linear relationship with antenna distance when the distance is between 1.5 and 4 m inside the laboratory (as shown in Fig. 4.13).

In these figures, transmission loss is well explained using approximate linear equations.

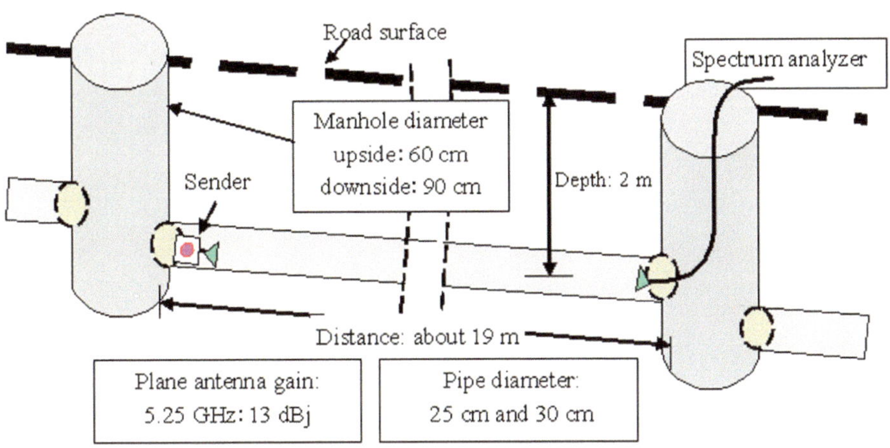

Fig. 4.10 System for the measurement of transmission loss

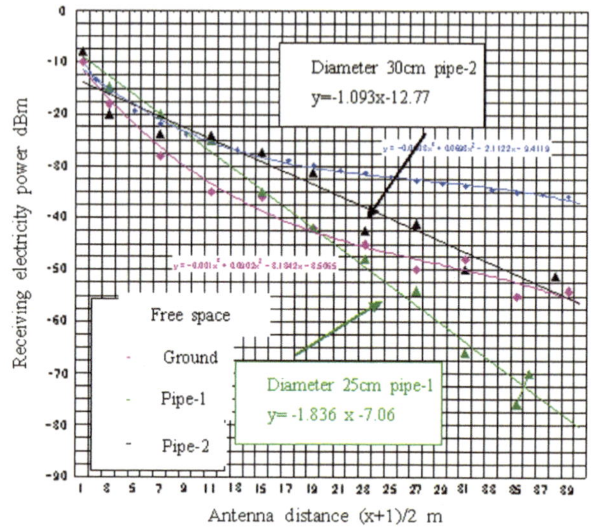

Fig. 4.11 Electric power output

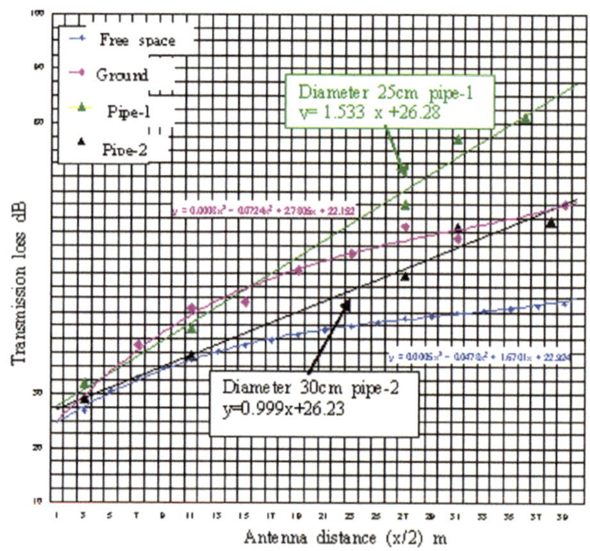

Fig. 4.12 Transmission property

4.3.4 *Discussion of Results*

Based on the ground transmission properties, the measurement results at close range, and the error, an approximate equation could be obtained to predict the transmission loss in the pipe, as shown in Fig. 4.12. The corresponding measured data are shown in Table 4.1.

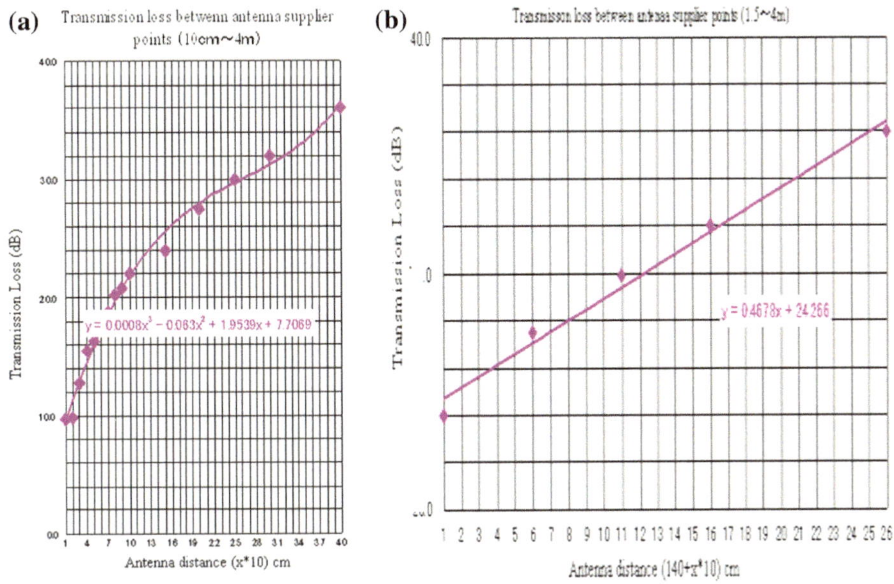

Fig. 4.13 Transmission loss between antenna and the transmitter in indoor condition. **a** Antenna and the transmitter distance 10 cm–4 m. **b** Antenna and the transmitter distance 1.5 m–4 m

Table 4.1 Approximate transmission loss in ceramic drain pipe

Distance (m)	Transmission loss (dB)	
	Pipe with a diameter of 25 cm	Pipe with a diameter of 30 cm
1	27.8	27.2
2	30.9	29.2
5	40.1	35.2
10	55.5	45.2
20	86.2	65.2
30	116.9	85.2

Transmission loss in the drain pipes is explained by the equations given below: For the pipe 1, with a diameter of 25 cm, transmission loss L is

$$L = 3.07 \times d + 24.75 \tag{4.1}$$

For the pipe 2, with a diameter of 30 cm, transmission loss L is

$$L = 2.00 \times d + 25.22 \tag{4.2}$$

From these equations, it is found that the transmission loss (dB) is proportional to the pipe length, as shown in Fig. 4.14.

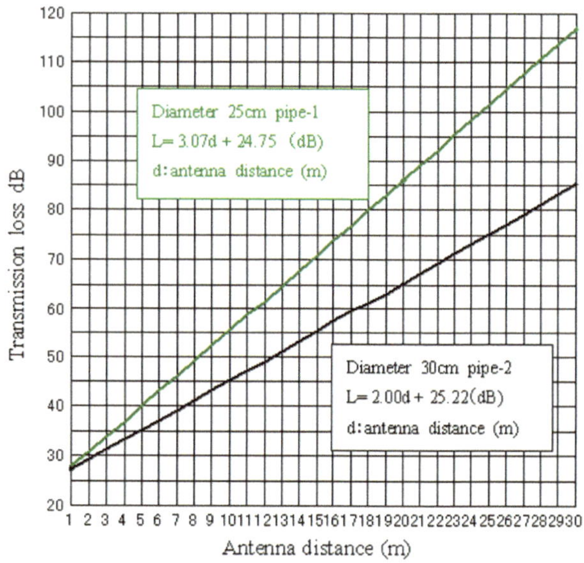

Fig. 4.14 Transmission loss between antenna electricity supplier points with antenna distance more than 1 m

If the antenna distance is more than 1.5 m, the transmission loss increases 3.07 dB/m for the 25-cm-diameter pipe and 2.0 dB/m for the 30-cm-diameter pipe, respectively.

4.4 Wireless Pipe Inspection System

The inspection robot system described above is used for indoor testing only. For real-life environment, a new inspection robot is needed. Therefore, a wireless communication system was introduced into a pipe inspection system for real-time data transmission and robot control.

4.4.1 Diagram of the Wireless Robot System

Figure 4.15 shows a diagram of the small wireless robot system.

The robot system consists of a PWM controlled motor drive system, VIA/CPU system, WLAN module and plane antenna system, light system, camera system, and 7.2 V battery power supply system.

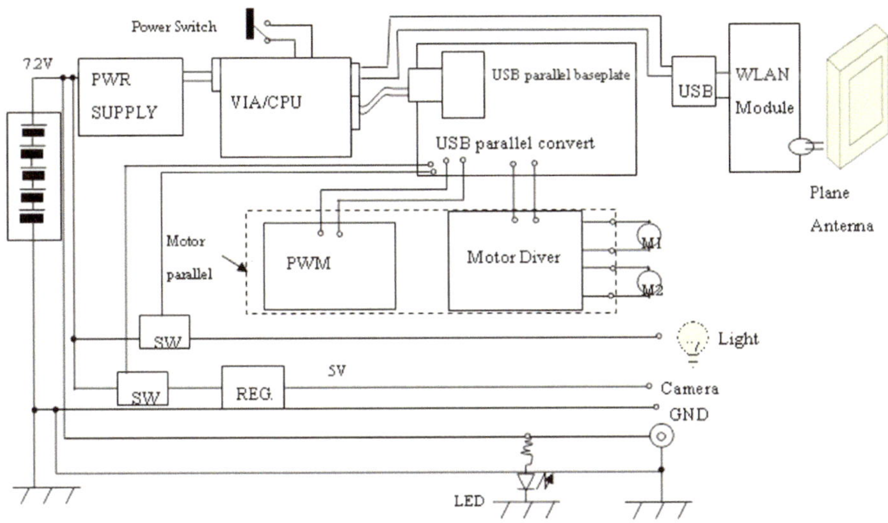

Fig. 4.15 Small-size wireless robot system

4.4.2 *Inspection Robot*

The inspection robot used is Mogurinko, developed by Ishikawa Iron Works, as shown in Fig. 4.16. This robot is used for pipes with a diameter of 25 cm.

The specifications of this inspection robot are listed below:

- Size: length 37 cm, width 18 cm, height 16 cm.
- Speed: 13.7 m/min.
- Driving mode: double motor.
- Electric Power: rechargeable 7.2 V batteries.
- Wireless frequency: 2.4/5 GHz.
- Data transmission: 100 base-T Ethernet.
- USB camera: 2,000,000 pixels.

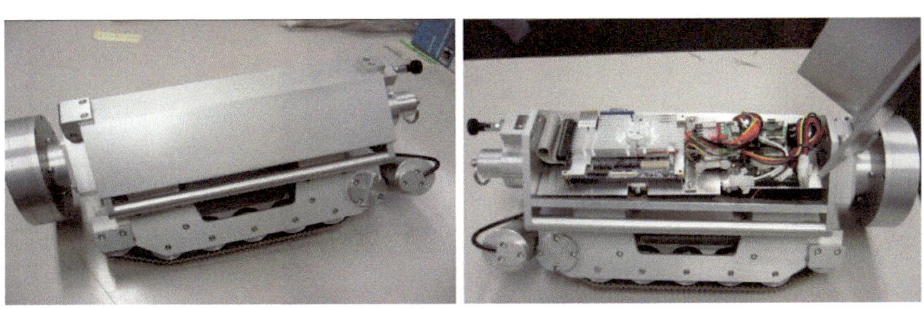

Fig. 4.16 Inspection robot (Mogurinko) and its interior

4.5 Platoon Moving

In this research, we have used platoon moving method to improve the poor wireless communication inside drain pipe environment. As shown in Fig. 4.17, we prepare two drain pipe robots, one robot is for drain pipe inspection mission, another one is used as an intermediate station to extend the wireless signal and keep a proper distance between two robots.

In order to keep the proper distance between two robots, the control system of follower robot is different than that of inspection robot. The control system of follower robot includes a H8 Microcontroller, an IR sensor, and a wireless network router (see Fig. 4.18).

Here, we adopt an IR sensor (see Fig. 4.19) to detect the distance by transforming analog output voltage variation according to distance which is about 0.4 ~ 2.6 V for distance ranging from 10 to 80 cm. According to the analog output signal, we can use A/D converter of H8 Microcontroller to judge the distance between two robots. The H8 Microcontroller has 10-bit A/D resolution, which is enough for detecting what distance we need, and also we can easily define the distance between two robots for platoon moving.

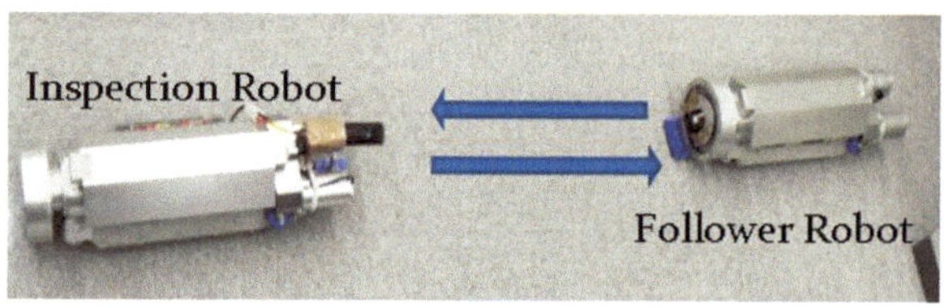

Fig. 4.17 Platoon moving

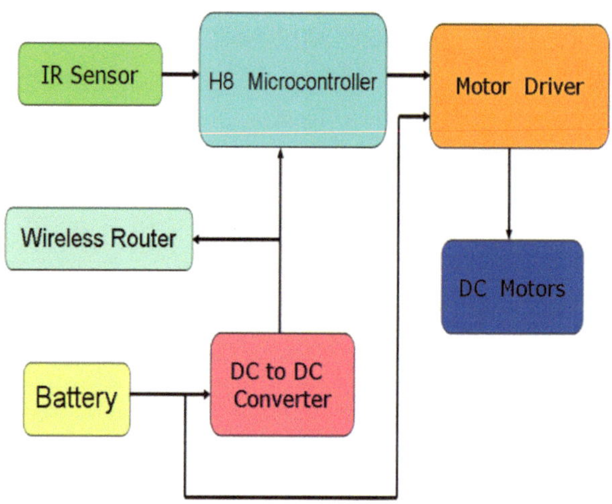

Fig. 4.18 Follower robot control system

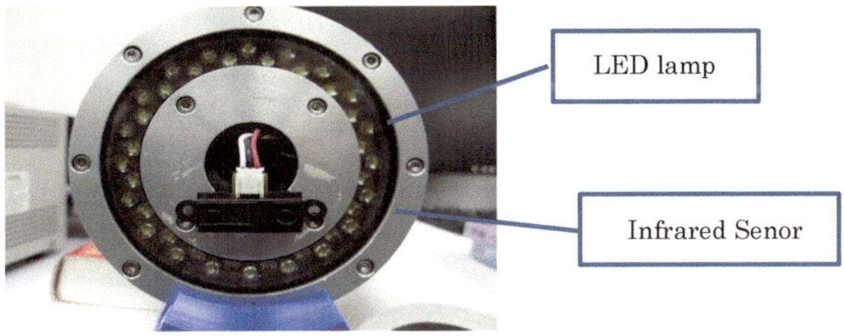

Fig. 4.19 IR sensor fixed in the robot

The follower robot also has an installed wireless network router as an intermediate station for extending the wireless signal range. It can achieve longer and stable operation of wireless communication inside pipeline between operator and drain pipe robot by the use of this device.

4.5.1 Experiment in Plastic Pipe

As shown in Fig. 4.20, we can control the drain pipe inspection robot movement into test plastic pipe with diameter of 30 cm and length of 30 m by wireless communication network.

Figure 4.20a shows the platoon moving action in plastic pipe. If the inspection robot (front robot) moves forward, then the follower robot (rear robot) will keep a proper distance between the two robots. Based on this method, we expect to use the follower robot to support more stable and longer distance for extending wireless network signal while moving into real environment of drain pipe.

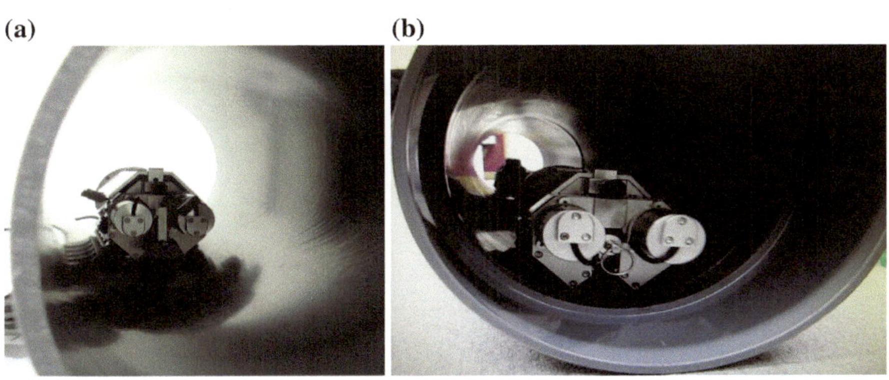

Fig. 4.20 Pipe inspection robot platoon moving test in plastic pipe

4.5.2 Experiment in Real-World Drain Pipe

As shown in Fig. 4.21, we tested wireless signal strength in real-world drain pipe test field. The drain pipe has width of about 24 cm and length of 20 m. The notebook using USB cable to connect wireless card is placed in the drain pipe inlet for wireless network signal strength analysis and control inspection (Fig. 4.22).

Figure 4.23 shows the follower robot's installed access point device as an intermediate station to extend the wireless signal between notebook (operator) and inspection robot (Fig. 4.24). The real-world drain pipe condition can be captured as high-quality image or videos form high-resolution CCD camera and save data into memory inside minicomputer.

Figure 4.25 shows the drain pipe image from inspection robot.

As shown in Fig. 4.26, we can see the variation of wireless signal strength from drain pipe inlet to exit. The wireless signal is approximately −20 dBm in the drain pipe inlet, and the wireless signal strength has decreased to −60 dBm due to the distance between the two robots.

Fig. 4.21 Real-world drain pipe test field [5, 6]

Fig. 4.22 Set wireless card in drain pipe inlet

Fig. 4.23 Installed access point in follower robot

Fig. 4.24 Inspection robot in real-world drain pipe

Fig. 4.25 Drain pipe image from inspection robot

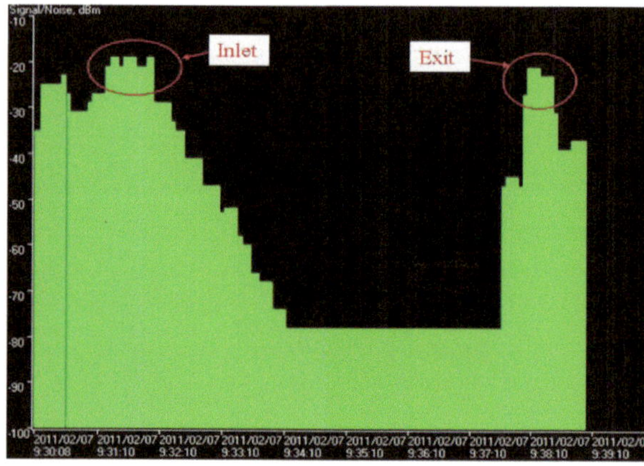

Fig. 4.26 Wireless signal strength between inspection robot and follower robot

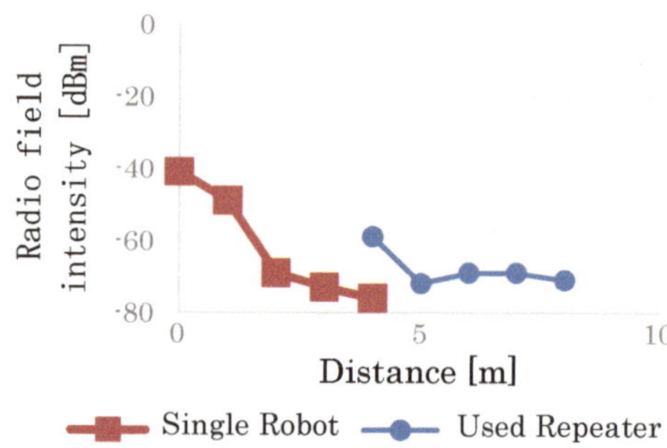

Fig. 4.27 Wireless radio field intensity by using repeater

Figure 4.27 shows another data using the first robot and second robot with repeater. From this data, it is clear that we can expand the transmission distance by using repeater.

4.6 Advanced Wireless Communication System

Two wireless communication systems are proposed for use in either a long pipe or a manhole. The type-1 wireless communication system is designed as shown in Fig. 4.28. In order to ensure the information transmission when passing through a manhole, a two-way amplifier between antennas has been introduced. Another type of wireless communication system is shown in Fig. 4.29. In this wireless system, leakage coaxial cable with a simple structure is embedded along the pipe to be tested. It is then easy to design a circuit for a wireless communication system. By integrating this system with an inspection robot, it is possible to test a longer pipe. It is predicted that this setup can allow information transmission for as much as 200–300 m. Moreover, with leakage coaxial cable, the robot's movement becomes easy

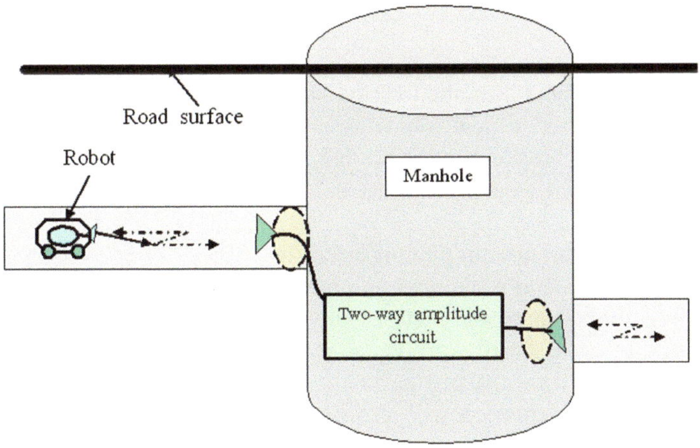

Fig. 4.28 Wireless communication system with two-way amplifier

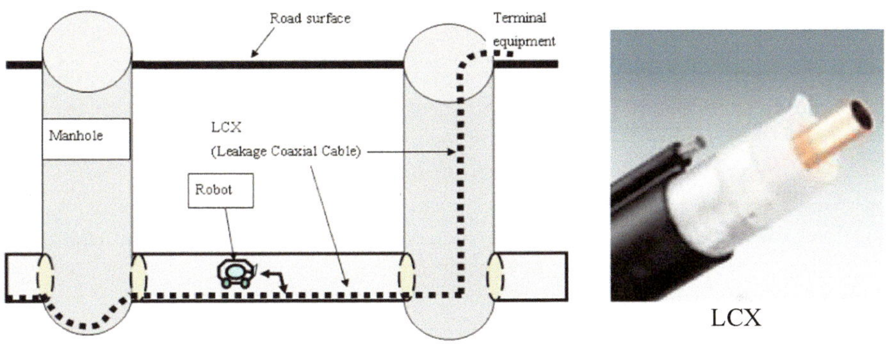

Fig. 4.29 Wireless communication system with leakage coaxial cable (LCX)

to control, and the reflection effect of the wall of the pipe would be reduced. In addition, this system can be applied to mesh cable and parallel two-wire cable.

The shapes of pipes can be very complex, and in reality, some manholes exist in a long pipe. In such a case, the two wireless communication systems should be integrated in order to compensate for the loss of data in the pipe.

Leaky coaxial cable has a slot of a constant period (based on the application frequency) on the outer conductor in order to radiate part of the signal energy as a radio wave to the space along the cable while simultaneously transmitting the signal. 'Antenna function' is a special structure coaxial cable. In the tunnel where the propagation characteristics of radio waves are poor, we use the LCX cable to realize the base station outside the tunnel to communicate with the mobile station radio inside.

4.7 Summary

In this chapter, we have described various testing of wireless transmission in steel and ceramic pipes, and a new robot inspection system for drain pipes integrated with wireless communication system has been discussed. Transmission losses inside two types of drainage pipes were measured, and transmission properties were obtained.

Owing to the differences in pipe materials and service conditions, the maximum transmission length of any wireless communication system should be determined individually. Furthermore, the state of any pipe is greatly affected by the environment. In future, the use of antenna and their specifications may be considered for smaller pipes, e.g., with a diameter smaller than 20 cm. The possible need for a higher frequency transmission will also be taken up in future research.

Model Questions

1. Compare the propagation characteristics of the radio waves of the steel tubes and ceramic tubes and describe the difference.
2. In the case of iron pipes, please explain why radio waves can propagate for long range.
3. From the Fig. 4.12, showing the loss characteristics of the radio wave in the ceramic tube, please compare the propagation characteristics of radio waves on the ground, in the air, and in the pipes of 25 and 30 cm diameters, and consider the difference and the explain the reason.
4. Describe how to propagate radio waves farther with ceramic tubes.
5. Describe what you must think, when you control and collect information by radio from an inspection robot inside a pipe.

References

1. Wada K (2007) Wireless radio communication system for pipe inspection robot and bridge diagnosis data. Waseda University, Master's thesis, pp 9–11
2. Ogai H, Wada K, Hirai K et al (2007) Wireless radio communication system for a pipe inspection robot. In ICASE International Conference 2007 (ICCAS 2007), Korea, 17–20 Oct 2007, pp 2616–2619
3. You W, Wu D, Ogai H, Hirai K, Abe T, Sato G (2008) Wireless radio communication system for a pipe inspection robot. In The 13th international symposium on artificial life and robotics. 31 Jan, 2 Feb 2008, pp 393–396
4. Ogai H, Yeh Y, Wu D, Hirai K, Abe T, Sato G (2009) Pipe inspection robot using wireless communication system. In The 14th international symposium on artificial life and robotics, 5–7 Feb, 2009, pp 59–62
5. Wu D, Ogai H, Yeh Y, Hirai K, Abe T, Sato G (2009) Pipe inspection robot using a wireless communication system. Artif Life Robot 14(2):154–159
6. Yang H, Oyabu R, Wang Z, Nishijima K, Ogai H, Pal N, Jain A (2011) Development of wireless communication and inside pipe image processing by using drain pipe robots with platoon moving. In The 11th SICE control division conference, Oinawa, March (2011) pp 1–6

Chapter 5
Design of Communication System and Corresponding Robot for Pipe Inspection

5.1 Introduction: Test Robot for Evaluating the Radio Wave Propagation Property

For factory plant piping, water pipelines, gas pipelines and the like, steel pipes are normally used. On the other hand, for sewer pipelines, ceramic pipes, Hume pipes (made of concrete) and PVC pipes (plastic) are generally used. The robots developed to inspect the condition of the pipelines capture the status of the inside of the pipes by the camera installed in them and transmit the image data to a base station that is placed in the tube inlet by a wireless communication system. In the last chapter, we have explained that for this system to communicate efficiently, in a steel pipe, a maximum of 100 m distance between the base stations and the robot has to be maintained for the diameter of the pipes between 20 and 40 cm. The transmission rate of the wireless communication should also be more than 10 Mbps. To meet these requirements, the wireless LAN of 2.4 GHz or 5.2 GHz should be used.

It is expected that in the case of steel pipes, due to multiple reflection in the conduit there will be a large propagation loss in the sewer pipe culvert. Hence, a thorough understanding of radio wave propagation properties in these pipelines is of primary importance. Therefore, to measure the radio wave propagation properties, steel indoor pipes of 30 cm diameter and 10 m length were installed in residential areas of IPS Waseda, Japan. Subsequently, for another test of wave propagation characteristics in steel pipe, out-door pipes of 20 cm diameter and 25 m length were set in the test bed at IITK, India. In this chapter, we will be discussing about the communication system and the robots used for this purpose.

H. Ogai and B. Bhattacharya, *Pipe Inspection Robots for Structural Health and Condition Monitoring*, Intelligent Systems, Control and Automation: Science and Engineering 89, https://doi.org/10.1007/978-81-322-3751-8_5

5.2 Wireless Communication System

In any such design problem, we will always begin with the outline of specifications based on which the system will be developed. The specification of wireless communication system with the inspection robot and the base station for steel pipe as well as ceramic pipe is as follows [1–4]:

- Two-way communication (full-duplex)
- Radio frequency (such as IMS band 5.6 GHz)
- Power capacity: DC power supply (5–10 V and 10 W)
- Communication data
  - Inspection Robot $\geq$ Base Station
    Moving images 640 × 480 30 frames/s transmission
    Static image 1024 × 1024 1 sheets/s (switched use)
    Position signal, such as 30 points 0.1 s cycle transmission
  - Base Station $\geq$ Inspection Robot
    Control signal 20 points 0.1 s cycle transmission
- Transmission speed 10 Mbps or more
- Wireless system IF Ethernet LAN for 10base-T socket
- Wireless system control signal (power on/off signal) by RS232C
- Antenna 100 mm × 100 mm within
  - Circuit within substrate size 100 mm × 100 mm

5.3 Design and Implementation of a Wireless Communication System

The wireless communication system has been designed as shown in Figs. 5.1 and 5.2. This system consists of a wireless communication base station connected with the PC. There are two modes of communication used for this purpose. RS232C is only used for power on/off switch, while the inspection data and the control signal are transmitted by the 10base-T-wired connection. 10base-T is an Ethernet-based wireless data transmission system, where '10' represents the communication speed of 10 Mbps, base represents base band signaling, and 'T' represents the twisted pair line. The base station has used a radio frequency of 5.6 GHz for communicating with the robot antenna. The robot is also connected with the antenna by a pair of RS232C and 10base-T system.

Here, we will discuss a case of sewer pipe robot radio design elaborately. To begin with, we consider the communicable distance in video transmission by wireless LAN. We select 5 GHz band wireless LAN with low attenuation based on measured value of radio wave attenuation in sewage pipe (diameters 25 and 30 cm).

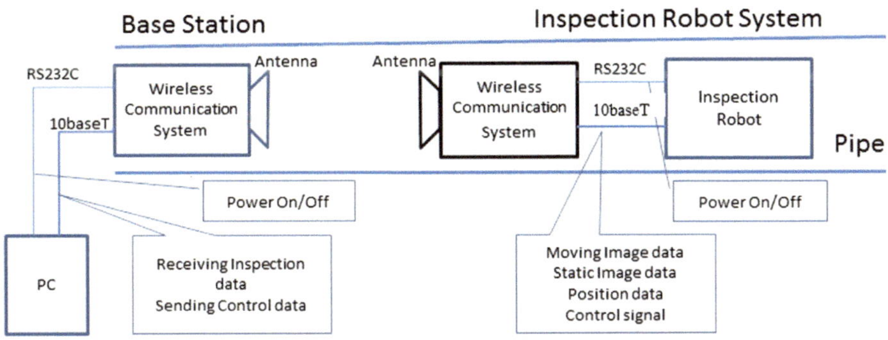

Fig. 5.1 Wireless communication system of the inspection robot inside the pipeline

Fig. 5.2 Wireless communication system

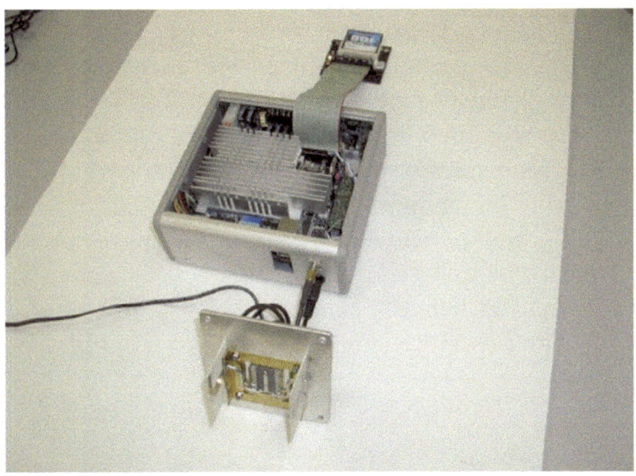

The criterion for determining whether video can be transmitted or not is indicated by **Eb/No.**

Eb is defined as the signal power per information bit—which is 77.3 dBm (0–10 log (54)—60) when the transmission power is 1 mW (0 dBm) and the transmission speed is 54 Mbps.

No is represented by noise power density—thermal noise power is defined as kTB (k: Boltzmann constant, T: absolute temperature, B: frequency bandwidth); accordingly No = −174.0 dBm/Hz.

The noise performance of the receiver is represented by a noise figure F, and the noise power (equivalent input noise power) converted to the input end of the receiver is FkTB. If the noise figure of the receiver is F = 4 (6 dB), the equivalent noise power is −168.0 dBm (−174 + 6).

The required value of Eb/No depends on the wireless modulation scheme. Orthogonal frequency division multiplexing (OFDM) is used in the 5 GHz band wireless LAN, and each subcarrier is subjected to 256 value of quadrature amplitude modulation (256 QAM). The theoretical value of Eb/No is 24.8 dB if the bit error rate that is practically required before the code error correction is set to 0.01%, and it is considered that Eb/No of about 26 dB is necessary for practical use.

Then, a signal power of −142.0 dBm (−168 + 26 = −142) is required at the wireless LAN receiver input terminal. The signal power input level per bit at the receiver is: transmitter output Eb (−77.3 dBm) + transmit antenna gain (Gt (dB)) − sewage pipe propagation loss (dB) + receive antenna gain (Gr(dB)).

The propagation loss in the sewer pipe allowed when Gt = Gr = 0 (dB) is 64.7 dB. From the experimental data on the propagation loss in the sewage pipe, the maximum transmissible distance is 13.4 m in the 25-cm-diameter pipe and 19.5 m in the 30-cm-diameter pipe.

The propagation loss in the sewer pipe allowed when Gt = Gr = 7 (dB) is 78.7 dB (the antenna gain of the experimental machine). It is 17.6 m with a 25-cm-diameter pipe and 26.2 m with a 30-cm-diameter pipe.

The maximum transmittable distance of an image can be increased by increasing the transmission power, reducing the number of transmitted images per second, or lowering the definition of the screen.

A prototype wireless communication system was built by using a 12 cm × 12 cm built-in processing board (VIA board made in Taiwan Pentium4, 1 GHz built-in board). The wireless communication prototype system is shown in Figs. 5.2 and 5.3. The battery system for this is shown in Fig. 5.4. It is generally placed under the robot.

Fig. 5.3 Wireless communication system

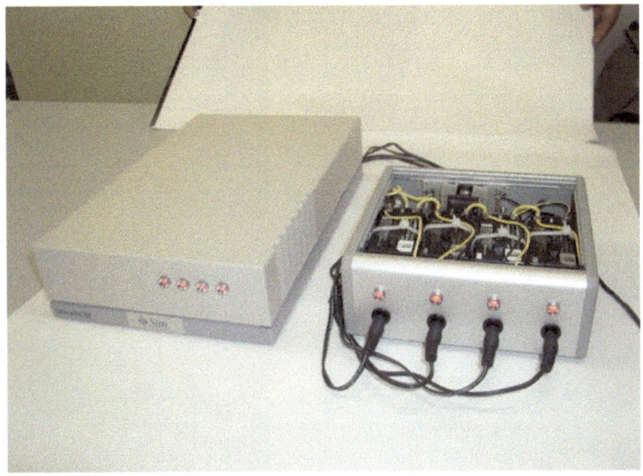

Fig. 5.4 Lithium battery power supply system and capacitor

5.4 Pipe Inspection Test Robot for Indoor Application

A Pipe inspection test robot that implements the wireless communication system is shown in Figs. 5.5 and 5.6.

The robot has two network cameras (Panasonic BB-HCM110) attached to its front, the wireless communication system attached at the top, the power system at the bottom and an antenna of 5 GHz at the rear.

The specification of the pipe inspection robot system is fixed as follows:

- Wireless communication system: 17 × 17 × 6 mm, Nano-ITX 10 W
- Power system: 7 × 17 × 6 mm
- Four 7.2 V 2.5 h lithium battery
- Carriage: 20 × 20 × 35 mm

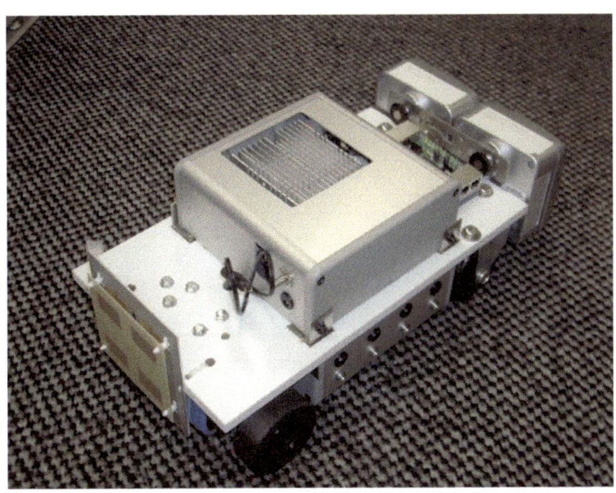

Fig. 5.5 Pipe inspection robot from the rear

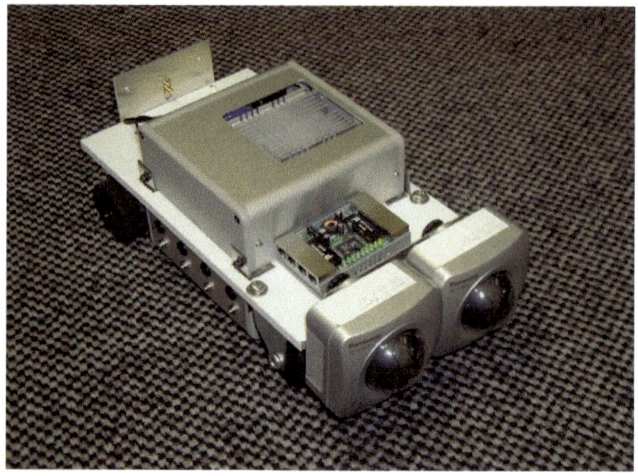

Fig. 5.6 Pipe inspection robot from the front

Table 5.1 Specification of camera (Panasonic BB-HCM110)

Image compression method	JPEG (motion JPEG in motion image display)
Resolution	640 × 480 dot, 320 × 240 dot, 160 × 120 dot
Image update speed (frame rate)	Up to 30 images/s (320 × 240 dot, 160 × 120 dot), up to 12 images/s (640 × 480 dot)
Zoom	10 times digital zoom
Angle of view	Left and right 43°, Up and down 32°
Pan (left and right)	±60° (remote operation)
Tilt (up and down)	+20° to −45° (remote operation)
Rotational speed	Pan: up to 80°/s, Tilt: up to 80°/s
Imaging element	1/4 inch, about 320 thousand pixels, CMOS sensor
Lens focal point	Fixed (focus range: 0.5 m to ∞)
Lens brightness	F2.8
Corresponding illuminance	5–10,000 lx
Temperature condition	Operating: 0–40 °C
External dimensions	(Width, height, depth) 100 × 100 × 73.5 mm
Mass (weight)	Approximately 275 g (body only)
Power supply	Dedicated AC adapter (AC 100 V 50/60 Hz, DC 12 V)
Power consumption	Standby state: about 2.5 W, maximum: about 5.5 W

- Geared motor × 2 (made by Tamiya)
- Two network cameras (forward)

Table 5.1 shows the specifications of the camera required.

Fig. 5.7 Pipe and robot in demonstration

5.5 Wireless Communication System Demonstration and Test in Steel Pipe

(1) Wireless Communication System Demonstration

Transmission capability is demonstrated by using this wireless communication system test as shown in Figs. 5.7 and 5.8, respectively. An experiment to transmit two network camera images and a DVD image stream is done by this system. It has been confirmed that the same can be transmitted without any hindrance.

(2) Wireless Communication Test in Steel Pipe

In an indoor laboratory test, the wireless communication test was done as shown in Figs. 5.9, 5.10 and 5.11.

Fig. 5.8 Wireless communication demonstration

Fig. 5.9 Wireless communication test robot

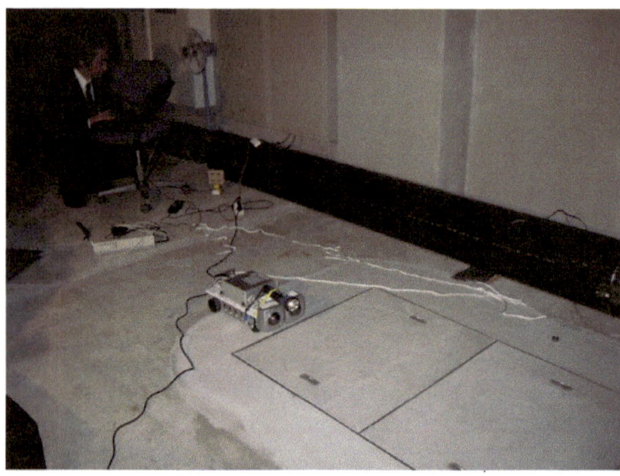

Fig. 5.10 Wireless communication test environment

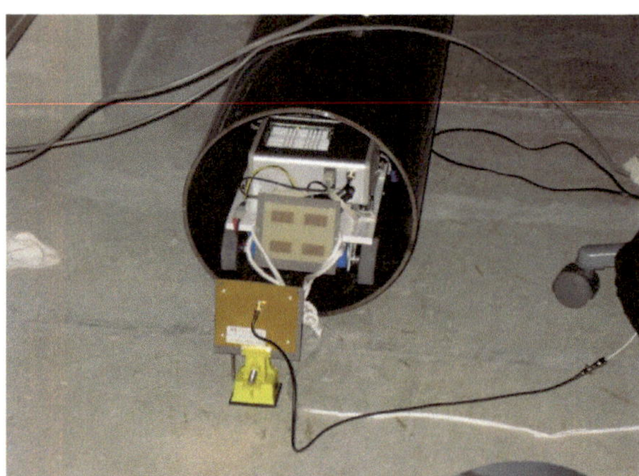

Fig. 5.11 Wireless communication test in steel pipe

Figure 5.9 shows the wireless communication test robot with a fluorescent light. Figure 5.10 shows the wireless communication test robot and the steel pipe. Figure 5.11 shows the test robot with wireless communication that enters the pipe.

As this figure suggests, the radio transmission test in steel pipe was conducted. The transmission result was shown and discussed in the previous Chap. 4.

5.6 Wireless Communication System Test for Drain Pipes

A sewer pipe is mainly a ceramic pipe. The wireless communication test by using a ceramic pipe was done as shown in Figs. 5.12 and 5.13 in the indoor laboratory.

Fig. 5.12 Ceramic pipe in the laboratory

Fig. 5.13 Wireless communication test inside the ceramic pipe

Figure 5.13 shows the connected ceramic pipes. Figure 5.14 shows the robot inside the ceramic pipe. This test results are also already discussed in the previous Chap. 4.

Figures 5.14 and 5.15 show the robot near the real drain pipe in Kitakyushu city, Japan. Figure 5.16 shows the interior of a ceramic pipe monitored by Web camera. Figure 5.17a shows the robot getting inserted into the pipe and Fig. 5.17b shows the removal of the same. This test results are also discussed in the previous Chap. 4 and hence will not be repeated here.

Fig. 5.14 Pipe inspection test robot (front)

Fig. 5.15 Pipe inspection robot (back)

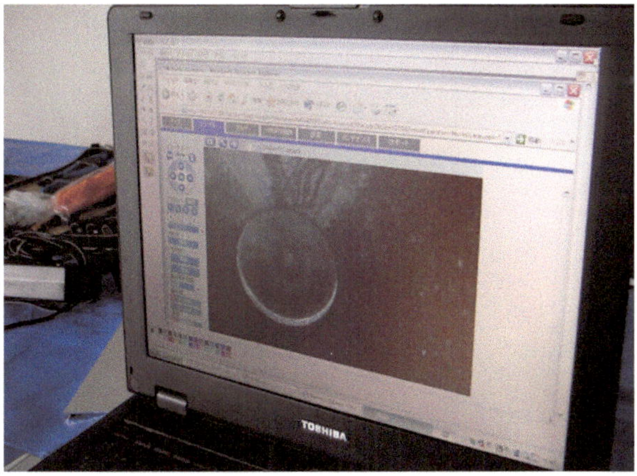

Fig. 5.16 Inside view of the monitored pipe as taken by the Web camera

(a) Inserting the Robot in the manhole **(b) Removal of the robot from the manhole**

Fig. 5.17 Wireless communication test inside a sewer pipe

5.7 Pipe Inspection Using Mogurinko 200

Another pipe Inspection robot is used subsequently which is customized from a commercial robot called Mogurinko 200, developed by Ishikawa Iron Works. The specification of the robot is shown in Table 5.2.

The robot architecture is shown in Fig. 5.18.

Figure 5.19 shows the system architecture for the wireless communication system along with the robot.

The specification of Small PC (Edison), shown in Fig. 5.20, is as follows,

- Power supply voltage DC 3.3–4.5 V
- Size 35.5 × 25.0 × 3.9 mm
- 500 MHz Intel Atom Silvermont dual-core processor

Table 5.2 Specification of MOGURINKO 200

Traveling speed	13.7 m/min
Drive system	Shaped arrangement crawler double motor
USB camera	CMS-010AFSV: max 1600 × 1200, max 30 fps
Power supply	7.2 V Ni-MH battery
Body dimension	L355 mm, W155 mm, H 150 mm 
Weight	5.4 kg

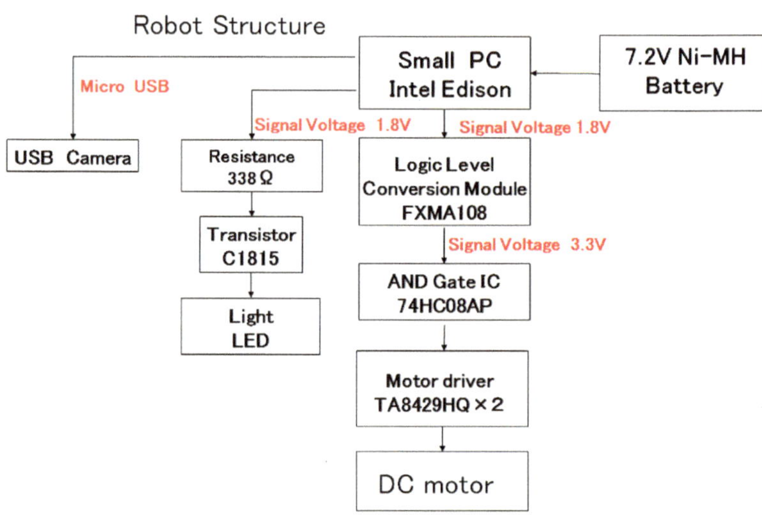

Fig. 5.18 Robot system structure

Fig. 5.19 System structure of wireless communication robot

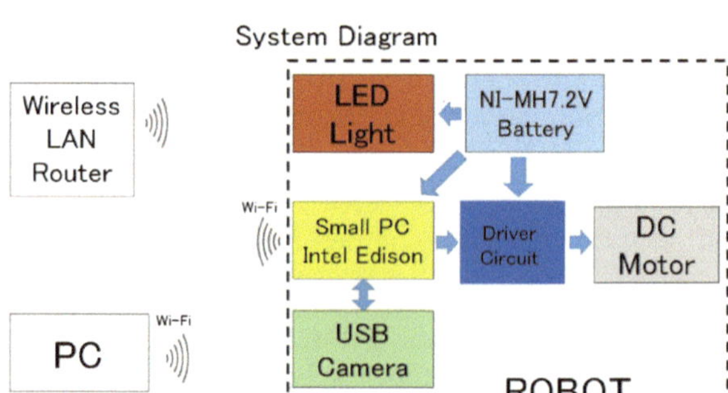

Fig. 5.20 Small PC Edison and breakout board

- RAM 1 GB DDR3
- Flash 4 GB eMMC
- Wi-Fi
- Bluetooth 4.0

And the specification of Edison Breakout Board is as follows:

- Board power supply DC 7–15 V
- Size 29 × 61.6 × 12.5 mm
- 0.1 Inch grid I/O array
- USB OTG (micro-USB TYPE-AB)
- USB–UART Bridge (micro-USB TYPE-B)

The specification of mobile Wi-Fi Router, shown in Fig. 5.21, is given in Table 5.3.

5.8 Advanced Pipe Inspection Robot Using Rotating Probe

A lot of pipe inspection robot that employs image processing technique has been developed up till now. However, whether the identified feature is a defect or a pattern might not be distinguished clearly only from the image processing. In such cases, the presence of the defect may be confirmed with a touch sensor. In this section, the configuration of such an advanced robot is described, which is integrated with the image processing sensor as well as a series of touch sensor on rotating probes, and the outcome of an experiment with the new robot are explained. Subsequently, we describe the future plans of integrating such probes.

First of all, continuous video recording inside the pipe is carried out as the robot passes through it. Next, the confirmation and the correction for the center point are applied as the images were taken by the camera from an asymmetric location. The final video image is made by filtering the image and correcting the seam. The recording power of the cameras of the image processing is 300,000 pixels CCD.

Fig. 5.21 Wi-Fi router (ZTE MF60)

Table 5.3 Specification of Wi-Fi router

Data service	HSPA+: DL 21.6 Mbps, UL 5.76 Mbps
	WCDMA CS: 64 kbps UL/DL
	WCDMA PS: 384 kbps UL/DL
Interface	USB and Wi-Fi
Wi-Fi standard	802.11 b/g/n
Wi-Fi security	WEP/WPA/WPA2/WAPI
USB version	USB 2.0 HIGH SPEED
USB interface type	Micro-USB
Weight	77 g
Dimensions (W × H × D):	99.7 mm * 53.7 mm * 13.9 mm
microSD card	Support up to 32 GB micro-SD card
Power adapter	5 V/700 mA
Battery	1500 mAh
Standby time	100 h
Working time	4 h

Next, the probe (sensor) that enabled the defect detection is developed. The sensor is installed in the rotational mechanism of ‘Mogurinko 250’ by ‘Ishikawa Iron Works Ltd.’ In a separate endeavor, the Indian Institutes of Technology

Kanpur has developed a rotating probe using piezo-element for inspecting the inside of pipes with a touch sensing system. In what follows is a brief description of the robot and the probe and its operational characteristics. Finally, we will discuss the outcome of a simple experiment with the probe installed in the advanced robot while traversing inside a pipe.

5.8.1 Robot Structure

In this study, the robot called 'Mogurinko 250' developed by 'Ishikawa Iron Works Ltd' was used. This robot has CCD camera in the front, and the rotating probe was installed in the back of the robot. Figure 5.22 shows the full system.

This robot can be controlled by wireless radio communication inside the pipe and can also transmit image information of the inside of the pipe in real time.

Specifications of this inspection robot are as follows [5–8]:

- Size: length 370 mm, width 180 mm, height 160 mm.
- Moving speed: 13.7 m/min.
- Driving mode: double motor.
- Electric Power: rechargeable batteries 7.2 V.
- Wireless frequency: apply to 2.4/5 GHz and Data transmission by 100base-T Ethernet.
- USB Camera (300 thousand pixels).

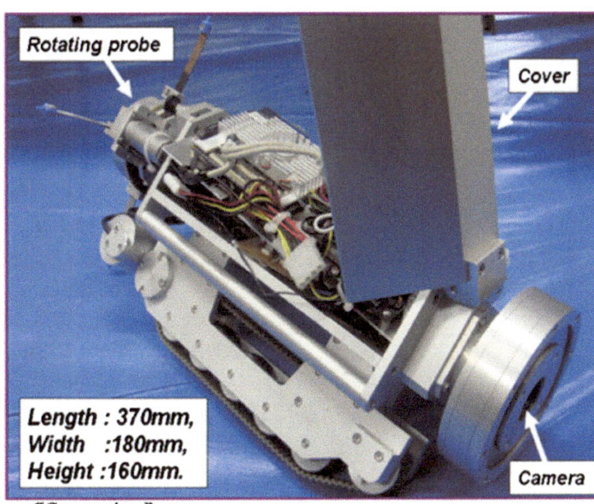

《Cooperation》
'Mogurinko250' by Ishikawa ironworks

Fig. 5.22 Mogurinko 250 fitted with the rotating probe

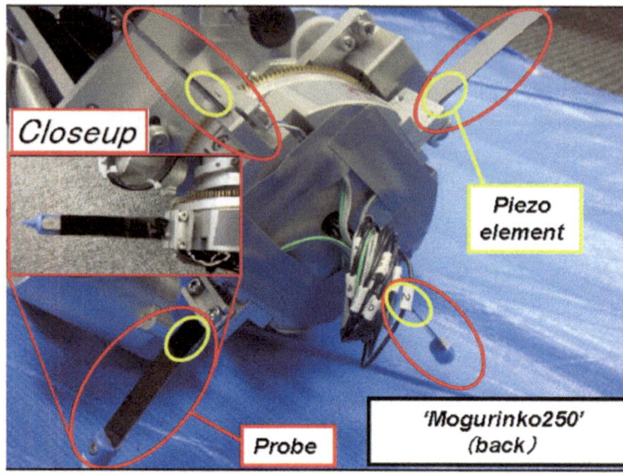

Fig. 5.23 Rotating probe

5.8.2 *Structure of the Rotating Probe*

The probe consists of spring steel with a piezo-film embedded at the base of the probe. The steel plate is used in cantilever condition. It is envisaged that during rotation, if there are obstacles, the probe will undergo an impact excitation which will be sensed by the piezo-film at the base. To house the cable of the rotating probe, a steel strip was used, as shown in Fig. 5.23 [9].

5.8.3 *Operating Principles of the Probe*

When the rotating probe touches the defected area, the piezo-film could detect the curvature change as there is a change of the dynamic stress. This stress change in the piezo-film can be measured as change of voltage. A typical motion sequence of the rotating probe is explained in Fig. 5.24.

First, the four probes rotate and approach the defect of the pipe surface (Fig. 5.24①). Next, the probes start to touch the defect and it detects the defect (Fig. 5.24②, ③). Afterward, the probes move away from the defect (Fig. 5.24④).

5.8.4 *Performance Study of the Probe Integrated with the Robot*

We begin the study with an experiment where we have used the robot in stationary condition, but the probes are in rotating mode inside a clean vinyl chloride pipe of 25 cm diameter, as shown in Fig. 5.25. The artificial defects were made of small strips of rubber\and fixed on the pipe surface.

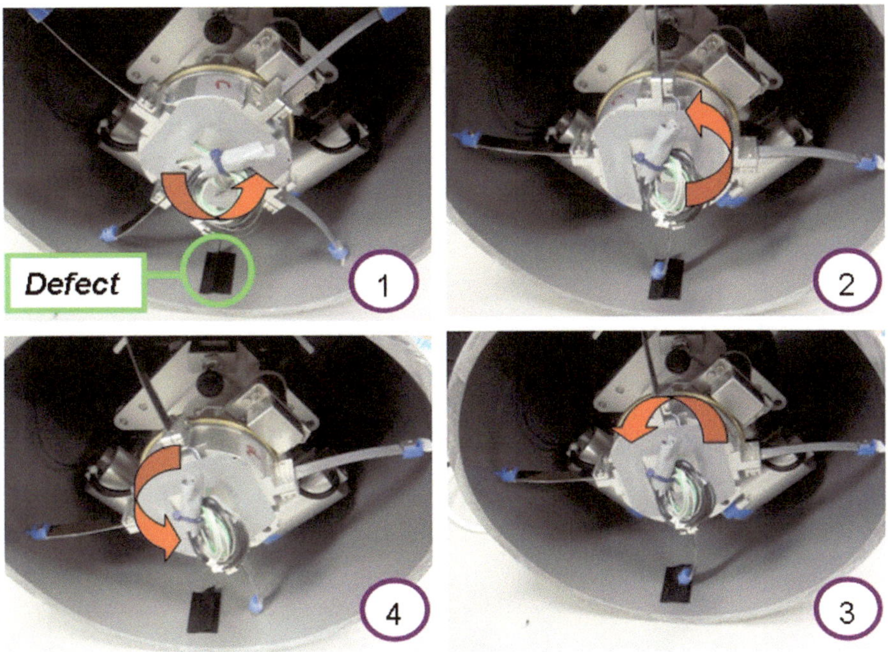

Fig. 5.24 Movement of rotating probe

Fig. 5.25 Experimental condition

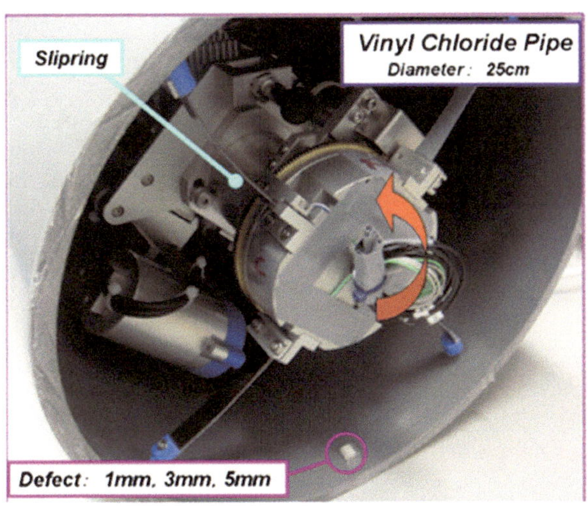

5.8.5 Data Capture Method

As explained earlier, the voltage change happens from bending of the probe. The voltage change signal is transferred to a microcomputer. It displays the same by using serial communications. The sampling period at this time is 0.01 s. The microcomputer used is H83664 tiny microcomputer. Figure 5.26 shows the microcomputer used for the application.

5.8.6 Details of Experimental Conditions

In all these experiments, the robot is kept stationary, while the rotating probe is designed to touch various experimental objects as detailed in Table 5.4. The resulting change of voltage is recorded and further analyzed to ascertain the detection capability of the probes.

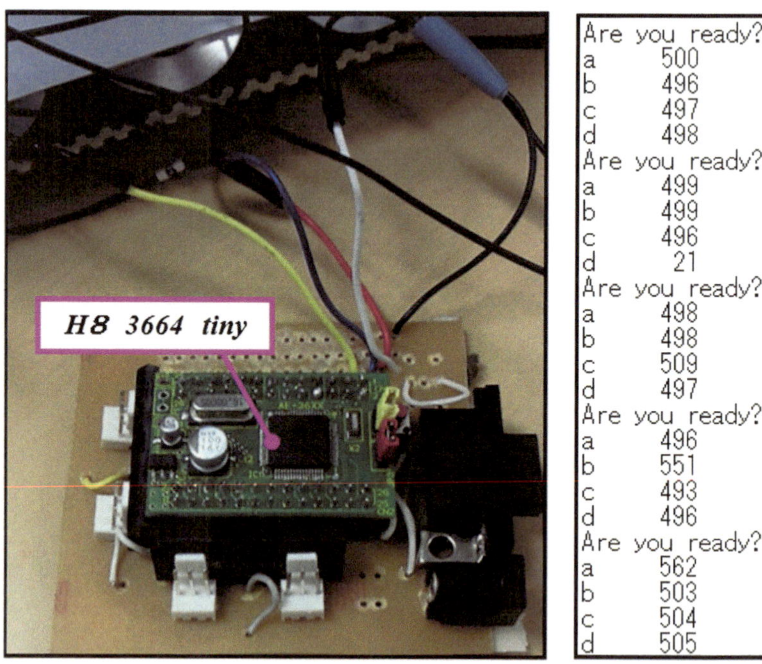

Fig. 5.26 Tiny microcomputer used for data logging

Table 5.4 Various experimental conditions

Experiment 1 (fundamental feature of probe)	Defect form: cube, cuboid (height: 10, 6, 3 mm, width: 10 mm)
Experiment 2 (detection of continuous defect)	Defect form: five cubes (height: 10 mm, width: 10 mm)
Experiment 3 (change on defect)	Defect form: three cuboids (cube) (height: 10 mm, width: 50, 30, 10 mm)
Experiment 4 (change according to form)	Defect form: Semicircular (diameter: 40, 30, 20 mm) defect form: equilateral triangle (sides: 20 mm)
Experiment 5 (damage experiment)	Defect form: concave type (width: 10, 6, 3 mm)

5.8.7 *Typical Experimental Results*

In order to exemplify, typical outcome of experiments 4 and 5 are shown in Figs. 5.27 and 5.28 respectively.

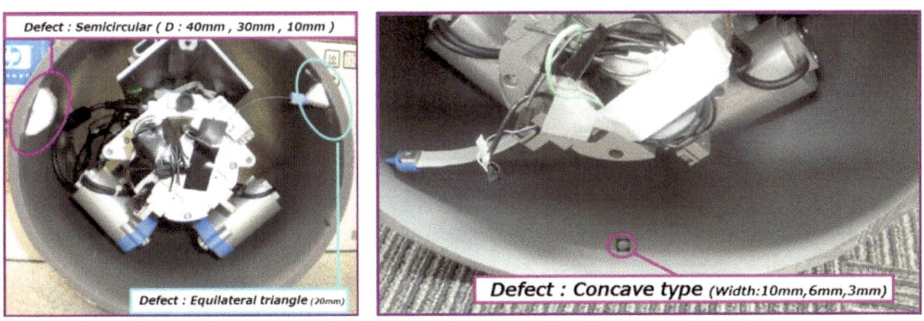

Fig. 5.27 Defect form: semicircular, equilateral triangle, and concave shape

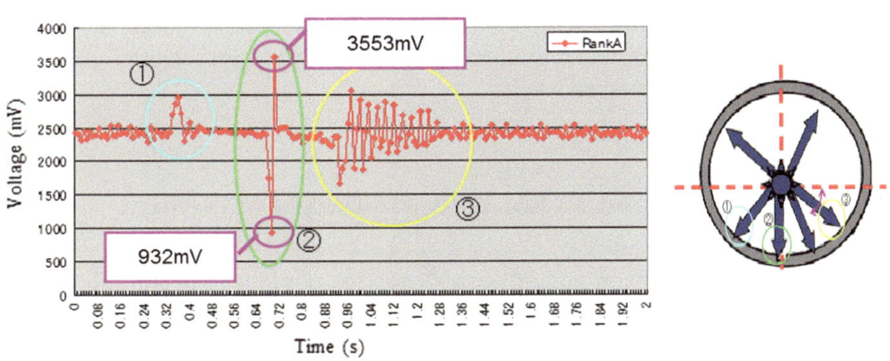

Fig. 5.28 Change of piezoelectric sensor voltage corresponding to Fig. 5.27

Table 5.5 Variation of voltage corresponding to various defect sizes

Condition of defect	Voltage (mV)
Defect type: cube (height: 10 mm)	More than 3,000
Defect type: cube (height: 6, 3 mm)	About 2,700
Defect type: concave (width: 10 mm)	About 900
Defect type: concave (width: 6 mm)	About 1,700
Defect type: concave (width: 3 mm)	About 1,900

Table 5.6 Experimental results (form)

Condition of defect	Voltage change
Defect type: cuboid (width: 50, 30, 10 mm)	• Irregularity on defect surface • width of defect
Defect type: semicircular (diameter: 40, 30, 20 mm)	• Smoothly increase and decreases
Defect type: equilateral triangle (sides: 20 mm)	• Fixed part and the tip of the probe (two times)

5.8.8 *Summary of Experimental Results*

Tables 5.5 and 5.6 show the outcome of the experiment in tabular form which clearly demonstrates the sensing potential of the rotating probe toward various sizes and shapes of defects.

5.8.9 *Automatic Defect Recognition Method*

In order to develop an algorithm such that the presence of defects can be directly predicted by the rotating probe, an automatic defect recognition method is developed based on these outcomes of experiments. Figure 5.29 shows a simple algorithm of automatic defect recognition method. In subsequent chapters, we have discussed how you can make a more complex detection algorithm by employing machine learning techniques.

5.9 Pipe Inspection Robot with Multi-axes Movement

For an effective pipe inspection mission, the design of pipe inspection robot should include some elements as noted below:

- Embedded power
- Friendly graphic user interface

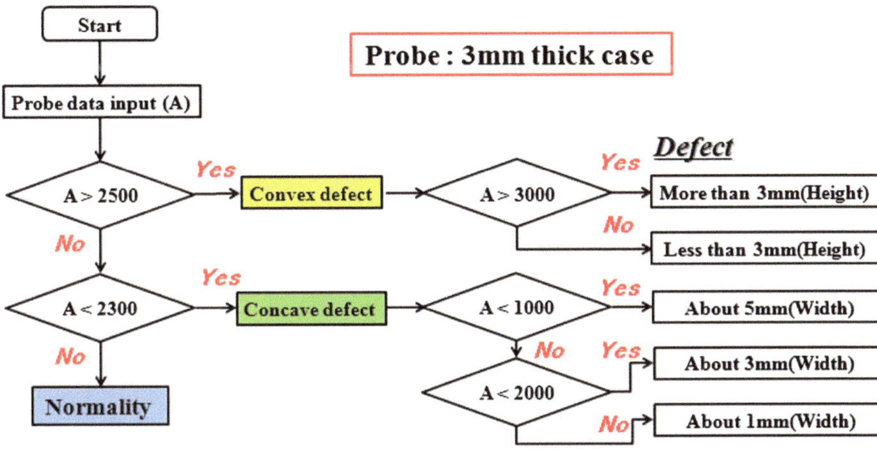

Fig. 5.29 A simple defect detection algorithm

- High resolution camera for damage detection
- Robust mechanical structure of robot
- Reliable wireless communication system
- Flexible movements for various pipes

Here we present a new pipe inspection robot with flexible movements as shown in Fig. 5.30. This robot has flexible mechanism designed for autonomous fit to the pipes within a diameter range of 230–250 mm. Through a friendly graphic user interface on operator pc, one can easily control the robot motion to perform in-pipe inspection mission by wireless network (Table 5.7).

Fig. 5.30 A flat chassis pipe inspection robot

Table 5.7 Specification of flexible pipe robot [9]

Specification	Value	Unit
Length	210	mm
Width	260	mm
Thickness	100	mm
Propelling force	4	kg
Weight	3	kg
Power consumption	30	W
Range of pipe diameter	230–250	mm

5.9.1 Control System

As shown in Fig. 5.31, we can remotely control the pipe inspection robot by wireless network. The NI-MH 7.2 V battery supports energy to power supply board and generates an output of 5 V for the mini-PC and control board. The CCD camera can capture high-quality picture or video and save data into mini-PC. The mini-PC can communicate with control board by RS-232 such that the control board output (in the form of PWM signal) can be used successfully to drive each of the servo motors. We can easily control the robot's motion in the pipe having diameters of 230–250 mm.

5.9.2 Experiments

(a) Horizontal Pipe Experiment

Figure 5.32 shows the motion sequence of pipe inspection robot in a horizontal pipe. First, we turn on the robot power and select the forward motion option. Then, we insert this robot into the transparent pipe inlet. Consequently, the robot goes straight and smoothly in the horizontal pipe.

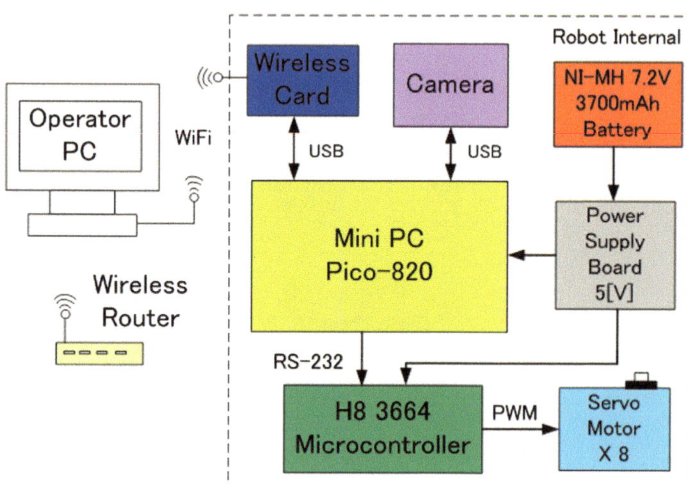

Fig. 5.31 Robot control system

Fig. 5.32 Motion sequence in horizontal pipe

(b) Rotation of Pivot for Multi-Axis Motion

In a complex pipe structure, the robot may need to change posture for adapting any kind of pipe direction. As shown in Fig. 5.33, this robot can move around the pipe axis and control the rotation angle we want. The sequence of motion is depicted in the image. First, we control the robot movement into the pipe. Second, we select rotational motion option and the steering servo motor turns the propelling wheels to the perpendicular direction. Third, we drive those propelling wheels to the same direction and the robot body starts to rotate around the pipe axis. Fourth, we stop all propelling wheels, the steering angle turns to 0°, and the robot comeback to the initial position.

(c) Vertical Pipe Experiment

In this experiment, the capability of driving in a vertical pipe is considered as the most important issue for mechanical design. As shown in Fig. 5.34, this experiment proved that this robot has enough propelling force and grip force between wheels and pipe wall, and the robot can successfully rise in a vertical pipe by the embedded power system.

(d) Helix Motion Experiment

Here we present a screw-like motion for saving energy while driving in a vertical pipe. As shown in Fig. 5.35, first the steering angle of servo motors is turned to 45°, respectively. Second, we drive those propelling wheels to the same direction and the robot body starts to rotate around the pipe axis. Third, the robot still rotates and rises in the vertical pipe in a helical motion. This is considered to be energy efficient as the rotation is coupled with forward movement.

5.10 Summary

In this chapter, we have explained the design specifications of wireless communication system and a range of pipe inspection test robots. We have tested the idea of wireless transmission system in a steel pipe and a ceramic pipe, respectively. Owing to the differences in pipe materials and service conditions, the maximum transmission length of any wireless communication system should be determined. Furthermore, the state of any pipe is greatly affected by the environment. Subsequently, we have discussed the issue of limitation of image processing-based defect detection. To enhance our sensing capability, we have discussed the development of a rotating probe sensing system based on piezoelectric material and shown its performance in a vinyl chloride pipe. In this chapter, we have also explained a pipe inspection robot with flexible movement. The robot has great

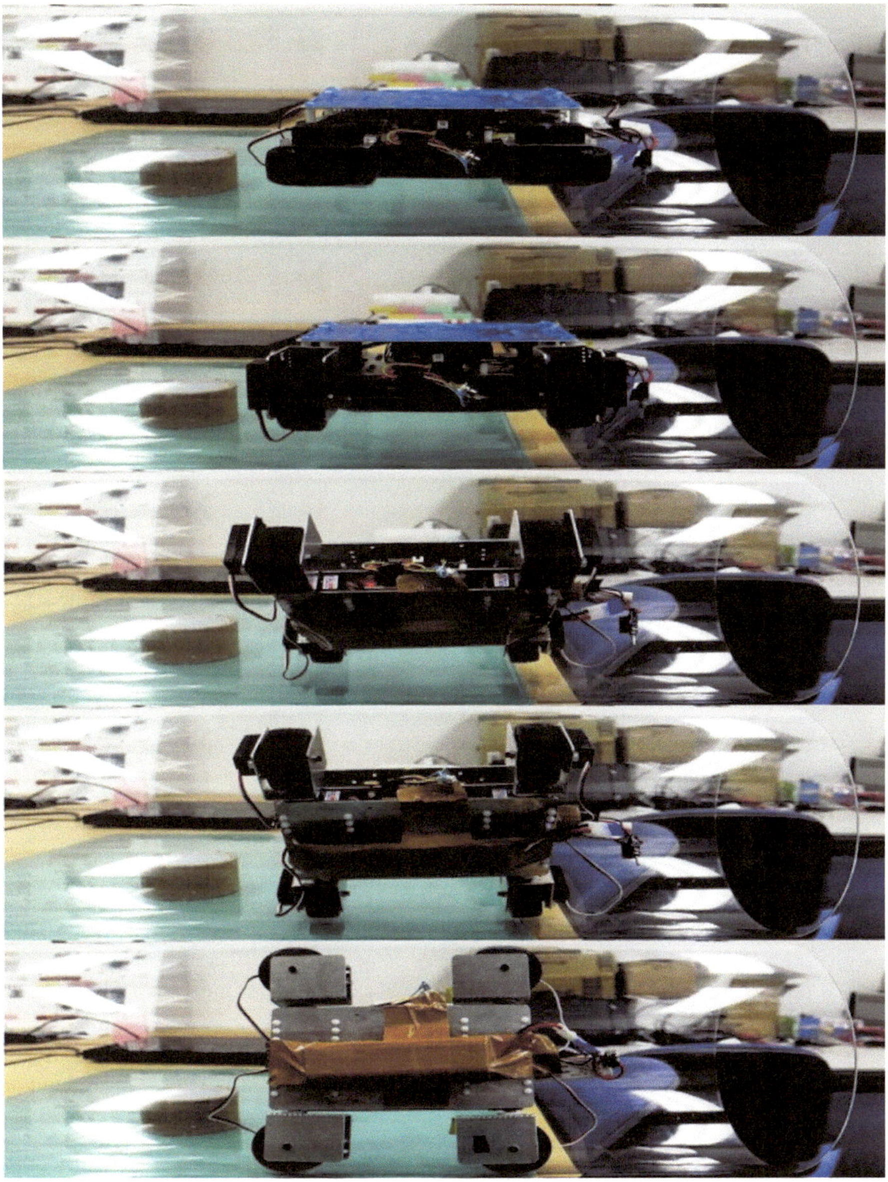

Fig. 5.33 Multi-axis rotational motion in a horizontal pipe

Fig. 5.34 Action in a vertical pipe

Fig. 5.35 Helix motion in a vertical pipe

mobility for driving into complex pipe structure. The experimental result shows that this robot can move into horizontal and vertical pipe smoothly. In addition, this robot can move around pipe axis and control whatever rotational angle we want. One can also generate a helical motion which is considered to be energy efficient during vertical motion.

Model Questions

1. Explain the components of an wireless communication system.
2. Explain the method of calculating the image transmission distance by wireless communication.

3. Explain the pipe inspection robot shown in Fig. 5.22 describing various components.
4. Explain the structure of rotating probe and its function.
5. Explain how to detect the defect of the pipe by using the rotating probe.
6. Explain the kinematics of the pipe inspection robot with multi-axes movement.

References

1. Ogai H, Wada K, Hirai K et al (2007) Wireless radio communication system for a pipe inspection robot. In: ICASE international conference on 2007 (ICCAS 2007), Korea, 17–20 Oct 2007, pp 2616–2619
2. You W, Wu D, Ogai H, Hirai K, Abe T, Sato G (2008) Wireless radio communication system for a pipe inspection robot. In: The 13th international symposium on artificial life and robotics, Jan 31–Feb 2, pp 393–396
3. Ogai H, Yeh Y, Wu D, Hirai K, Abe T, Sato G (2009) Pipe inspection robot using wireless communication system. In: The 14th international symposium on artificial life and robotics, 5–7 Feb 2009, pp 59–62
4. Ogai H, Wang Z, Takeno S (2010) Real-time generation of developed view for drain pipe based on camera video. In: The 15th international symposium on artificial life and robotics, 4–6 Feb 2010, pp 581–584, 29–53
5. Nassiraei AAF, Kawamura Y, Ahrary AA, Mikuriya Y, Ishii K (2007) Concept and design of a fully autonomous sewer pipe inspection mobile robot KANTARO. In: Proceedings of IEEE international conference on robots and automation, Apr 2007, pp 136–143
6. Lu C-P, Huang H-P, Yan J-L, Cheng T-H (2007) Development of a pipe inspection robot. In: The 33rd annual conference of IEEE industrial electronics society, 5–8 Nov 2007, Taipei, Taiwan, pp 626–631
7. Fjerdingen SA, Liljebäck P, Transeth AA (2009) Proceedings of IEEE/RSJ international conference on intelligent robots and systems, 11–15 Oct 2009, St. Louis, USA, pp 5665-5
8. Wang Z, Ogai H, Takahiro S (2010) Real-time generation of developed view for sewer pipe based on web camera video. In: The 15th international symposium on artificial life and robotics 2010, CD-ROM papers, OS3-4, Beppu, Japan
9. Nishijima K, Sun Y, Srivastava RK, Ogai H, Bhattacharya B (2010) Advanced pipe inspection robot using rotating probe. In: The 15th international symposium on artificial life and robotics 2010, CD-ROM papers, OS3-2, Beppu, Japan
10. Yang H-P (2013) Development of the pipe inspection robot with flexible movements. Thesis of Master of Engineering, Waseda University

Chapter 6
Video-Based Sensing and Image Processing of Sewer Pipes

6.1 Introduction

With the rapid increase of the number of sewer pipes which have crossed their designed life, the sewer pipes are getting decrepit. On the other hand, the requirements of rebuilding the sewer pipes and the laying of optical fiber in the sewer pipes are increasing. Hence, in order to estimate the health of the sewer pipe accurately, frequent examination of the sewer pipes is becoming more and more important [1–5].

In Japan, from 1982, we have used robot-inspections with video cameras instead of human inspection for small-bore sewer pipes that cannot be accessed directly. Furthermore, in order to understand the pipe condition visually, systems which can generate the developed view of sewer pipes have been produced and been put into practical use. However, systems being used now usually create the unfolded view once the pipe inspection process is over, and hence, it is time-consuming and less effective. Because of this problem, we cannot confirm the condition of sewer pipes during investigation. The algorithm, we will present in this chapter, uses the computer vision library OpenCV and can provide a faster generation of the developed view which is possible to be processed in real time. Also, the algorithm can automatically identify the center points of different types of sewer pipes. Therefore, it enables us to check and record the developed view of the sewer pipes during the investigation itself, which results in significant improvement of the accuracy and efficiency of the pipe inspection. As a further extension, we plan to develop an auto visual inspection system to recognize the defects based on this algorithm together with some auto-detection algorithms. In this chapter, we first introduce the wireless inspection robot that we have used in our research briefly and then describe the unfolding algorithm and the center identification algorithm at length. The real data taken by our robot will be introduced and used during the

H. Ogai and B. Bhattacharya, *Pipe Inspection Robots for Structural Health and Condition Monitoring*, Intelligent Systems, Control and Automation: Science and Engineering 89, https://doi.org/10.1007/978-81-322-3751-8_6

discussion. It may be noted that although we have used a specific example of sewer pipe, the concept of image processing is generic and can be applied to other transportation pipes as well.

6.2 Hardware Environment

The inspection robot we have used in this research is developed together with the Technical Solutions Company ISHIKAWA IRON WORKS, which is shown in Fig. 6.1. The robot has the following specifications (Table 6.1) [6].

Comparing with the former inspection robots, the new robot has the following modified features.

(a) Equipped with wireless device, we can control and get the sewer image in the ground using a terminal computer instead of using cables like the former ones. This also makes it possible for us to do some processing of the real-time videos obtained through the wireless device.
(b) The eight-shaped crawlers enable robot-inspector to run more stably in the sewer pipe, which provides more smooth motion for the image processing as it can reduce the errors caused by the shaking of the camera.

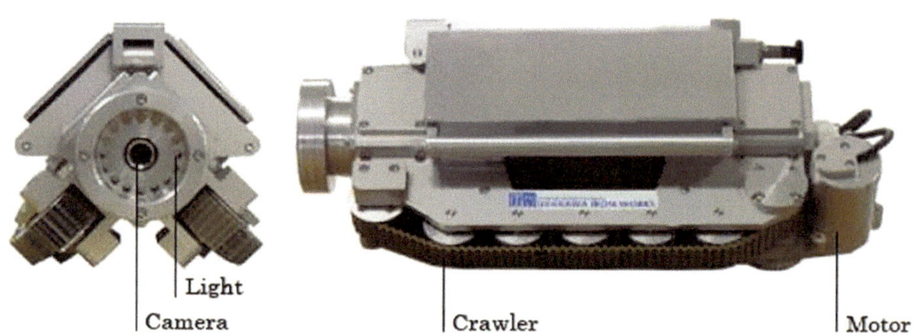

Fig. 6.1 Structure of the inspection robot

Table 6.1 Specifications of robot

Speed	0.228 m per s or 0.82 km/h
Camera	44 megapixels CCD
Drive system	Double motors/eight-shaped crawlers
Power supply	Two 7.2 V batteries
Pipe diameter	200 mm
Size	L 355 mm × W 155 mm × H 150 mm/5.4 kg

In addition, we are also trying to improve this robot by integrating with other sensors to recognize the defects. For example, a set of rotating touch-probe arms is under study now.

6.3 The Unfolding Algorithm

Unfolding algorithm is the process to generate the developed view (also unfolded view) of the sewer pipe. In order to make the developed map, the following processes should be executed, with reference to Fig. 6.2.

(a) Unfolding process: to open out the circular image and lay it flat to a band one.
(b) Connection process: to connect the unfolded images together from several frames to generate a developed map of the sewer pipe.

Here, we first introduce the unfolding algorithm.

The unfolding process is used to generate the unfolded image of the sewer pipe. This can be finished by two steps, coordinate transformation and interpolation. The coordinate transformation is used to calculate the position of point of the unfolded image in the original one, while the interpolation step is used to estimate the value of this point according to the nearby values of the pixels. This process is shown in Fig. 6.3.

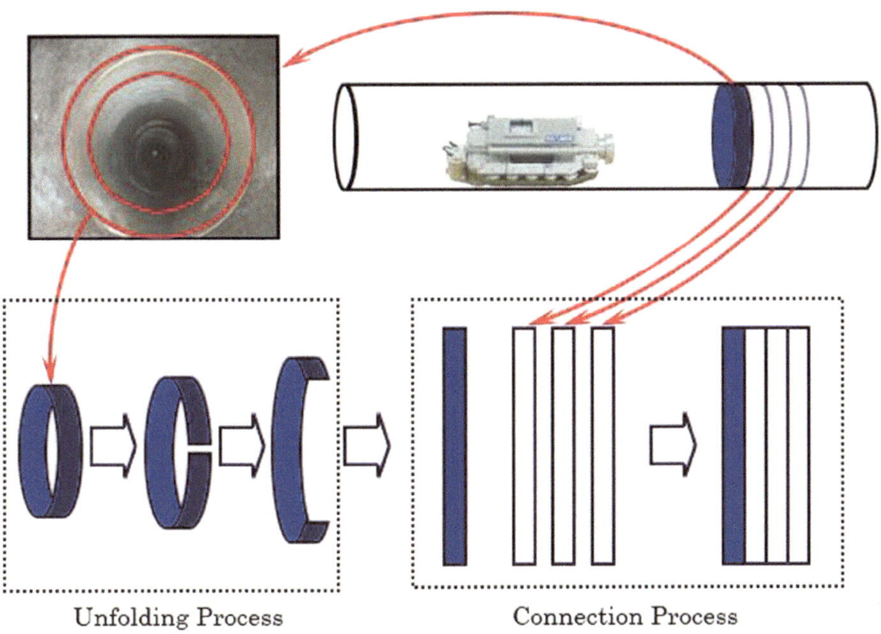

Fig. 6.2 Processes for developed view

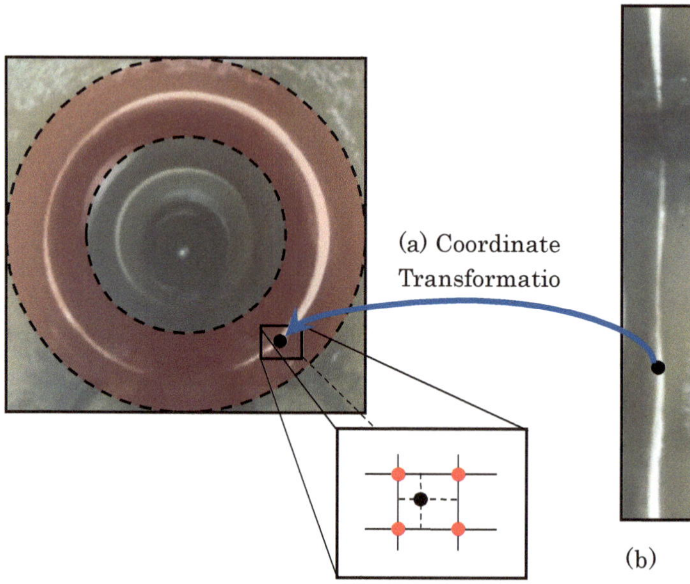

Fig. 6.3 Executing steps of the unfolding process

6.3.1 Coordinate Transformation

We can find that the original image is simply the polar form of the unfolded one. Hence, using the polar transformation together with some calculation, we can obtain the coordinate relationship between these two images.

For example, coordinate (x, y) denotes the position of a certain pixel point in the unfolded image and coordinate (x', y') denotes the related position of it in the original image (shown in Fig. 6.4). The size of the unfolded image is $M \times N$ and size of the original image is $M_0 \times N_0$, the unfolding region is from rMin to rMax, and the central point is (x_0, y_0) in the original image (Fig. 6.3). Then the transformation relationship can be calculated according to the following formula.

$$\begin{cases} x' = r\cos\theta + x_0 \\ y' = r\sin\theta + y_0 \end{cases} \quad \begin{cases} r = r\mathrm{Min} + \frac{r\mathrm{Max} - r\mathrm{Min}}{M} x \\ \theta = \frac{2\pi}{N} y \end{cases}$$

We can find that the x' and y' in the original image are both functions of the position (x, y). So we can generate two matrices to store the relationship of the coordinate (x, y) and (x', y'), for example, matrix $\boldsymbol{X}$ to store all the values of the x-coordinate x' and matrix $\boldsymbol{Y}$ to store all the values of the y-coordinate y'. Then, the position (x', y') can be denoted as $(\boldsymbol{X}(x, y), \boldsymbol{Y}(x, y))$. It is meaningful for us to generate these two matrices. First, we can use them as the inputs (map$x$ and map$y$) for the OpenCV function ***cvRemap()*** [7] which provides a fast transformation of

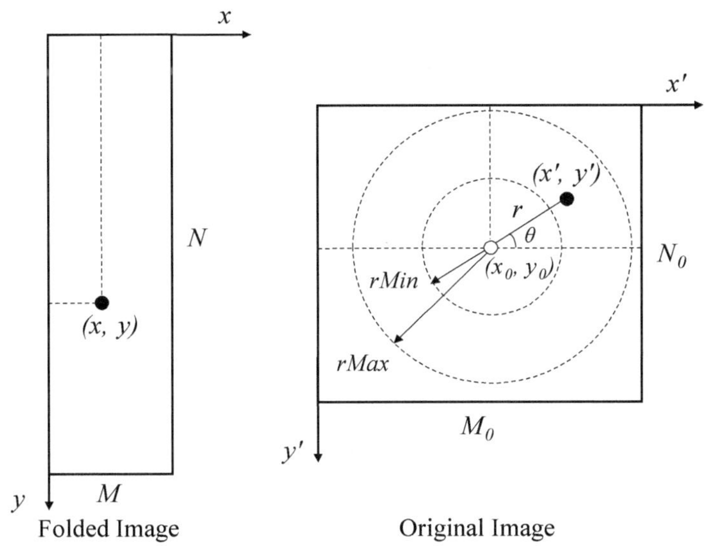

Fig. 6.4 Parameters for the coordinate transformation

two images with a certain interpolation method. Second, for the fault detection process, when we find the position of fault in the unfolded image and want to know the position in the original one, it will be convenient to use these two matrices rather than calculating it again. The related coding is given below.

```
void cvRemap( const CvArr* src, CvArr* dst, const CvArr* mapx, const CvArr* mapy,
              int flags=CV_INTER_LINEAR+CV_WARP_FILL_OUTLIERS,
              CvScalar fillval=cvScalarAll(0) );

    src ...................... Source image
    dst ...................... Destination image
    mapx .................. The map of x-coordinates
    mapy .................. The map of y-coordinates
    flags .................... A combination of interpolation method and the optional flag
    fillval ................... A value used to fill outliers

The function cvRemap transforms source image using the specified map:
              dst(x,y) <- src(mapx(x, y), mapy(x, y))
```

6.3.2 Interpolation

When an image needs to be scaled up, each pixel of the original image needs to be moved in a certain direction based on the scale constant. However, while scaling up an image by a nonintegral scale factor, there are pixels (i.e., *holes*) that are not assigned appropriate pixel values, whereas for the image all the pixel points are located at integer coordinates. In this case, those *holes* should be assigned appropriate RGB or grayscale values so that the output image does not have non-valued pixels. **Interpolation** can be used where perfect image transformation with pixel matching is impossible, so that one can calculate and assign appropriate intensity values to pixels. In our unfolding algorithm, we use the bilinear interpolation method as for its better balance of quality and computation speed (a discussion for different interpolation methods will be taken up later.

Bilinear interpolation [8] method is an interpolation method which uses four nearest pixel values located in the diagonal directions from a given pixel. The key idea is to perform linear interpolation first in one direction and then again in the other direction. An example of bilinear interpolation method is shown in Fig. 6.5.

Suppose that we want to find the value of the unknown function f at the point $P = (x, y)$. It is assumed that we know the value of f at the four points $Q_{11} = (x_1, y_1)$, $Q_{12} = (x_1, y_2)$, $Q_{21} = (x_2, y_1)$, and $Q_{22} = (x_2, y_2)$.

We first do linear interpolation in the x direction. This yields

$$f(R_1) \approx \frac{x_2 - x}{x_2 - x_1} f(Q_{11}) + \frac{x - x_1}{x_2 - x_1} f(Q_{21})$$

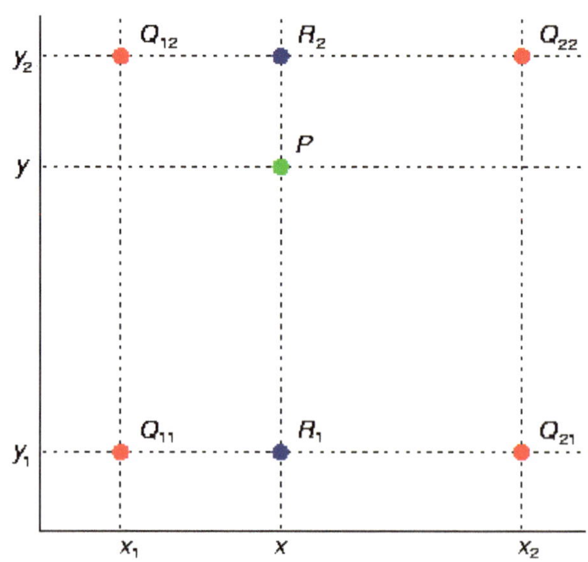

Fig. 6.5 Bilinear interpolation

where $R_1 = (x, y_1)$,

$$f(R_2) \approx \frac{x_2 - x}{x_2 - x_1} f(Q_{12}) + \frac{x - x_1}{x_2 - x_1} f(Q_{22})$$

where $R_2 = (x, y_2)$. We proceed by interpolating in the y direction.

$$f(P) \approx \frac{y_2 - y}{y_2 - y_1} f(R_1) + \frac{y - y_1}{y_2 - y_1} f(R_2)$$

This gives us the desired estimate of $f(x, y)$.

$$\begin{aligned} f(x, y) \approx & \frac{(x_2 - x)(y_2 - y)}{(x_2 - x_1)(y_2 - y_1)} f(Q_{11}) + \frac{(x - x_1)(y_2 - y)}{(x_2 - x_1)(y_2 - y_1)} f(Q_{21}) \\ & + \frac{(x_2 - x)(y - y_1)}{(x_2 - x_1)(y_2 - y_1)} f(Q_{12}) + \frac{(x - x_1)(y - y_1)}{(x_2 - x_1)(y_2 - y_1)} f(Q_{22}) \end{aligned}$$

It may be noted that the performance of the interpolation method is very important in our research as all the following processes, such as image connection and fault detection, are all based on the unfolded image. So it is necessary for us to choose a suitable interpolation method with good performance as well as high speed to meet the real-time requirement. The commonly used interpolation methods are **nearest neighbor interpolation**, **bilinear interpolation,** and **bicubic interpolation**. We will give a comparison of these three methods based on the computational complexity and the interpolation performance. The computational complexity can be seen from Table 6.2 [9].

Generally speaking, the performance of nearest neighbor interpolation is not satisfactory as it does no reduction of the visual distortion caused by resizing, even though it has the fastest speed. The bilinear and bicubic interpolation can give a reduction of this distortion. Also, as the bicubic interpolation uses a larger interpolation area, the performance of it is better than bilinear interpolation. So considering the computational complexity and interpolation performance, we choose the bilinear interpolation method in our unfolding algorithm.

Table 6.2 Computational complexity of interpolation methods

Interpolation method	Interpolation region	Times of multiplication	Times of addition	Others
Nearest neighbor interpolation	0	0	0	Just copy the nearby pixel, so there is no additional calculation
Bilinear interpolation	2 × 2	6	3	
Bicubic interpolation	4 × 4	20	15	

6.4 Hough Transform Method

The pipe-joints of the sewer can be obviously seen caused by the reflection of the light. Usually, the pipe-joints are almost of circular forms. So if we treat them as circles, we can use Hough transform to recognize them. Figure 6.6 shows the results by Hough transform we used in our algorithm.

Hough transform is a feature extraction technique used in image analysis to find features of a shape in an image by voting procedure. In theory, the general Hough transform can be used on any kind of shapes, although the complexity of the transformation increase with the number of parameters needed to describe the shape. In practice, it is only generally used for finding straight lines or circles.

The Hough transform can be described as a transformation of a point in the x, y plane to the parameter space. The parameter space is defined according to the shape of the object. For example, a circle can be described in the x, y plane by:

$$r^2 = (x - a)^2 + (y - b)^2 \tag{6.1}$$

where a and b are the center of the circle in the x and y direction, respectively, and r is the radius.

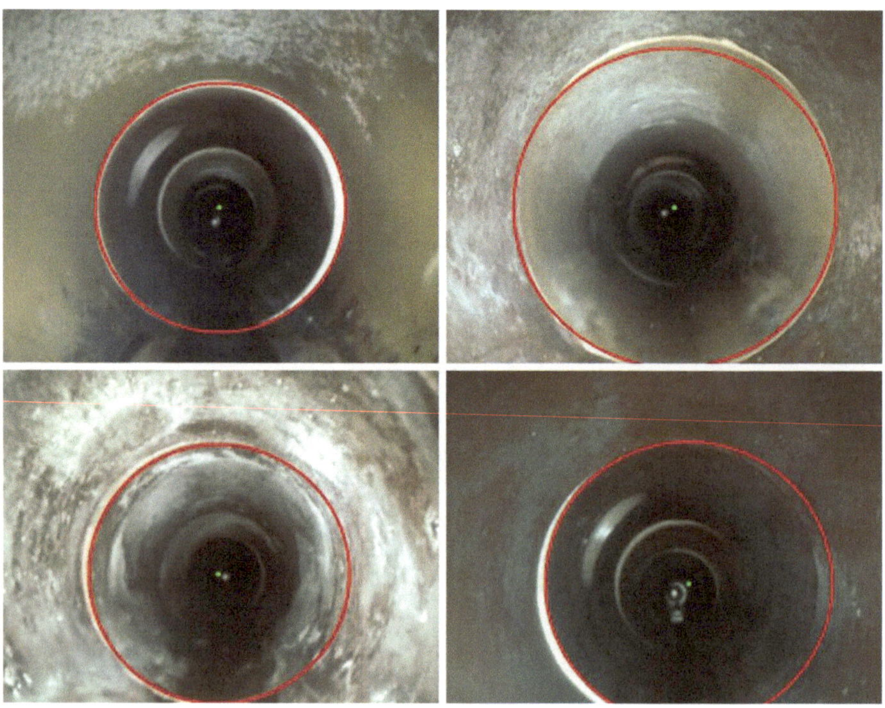

Fig. 6.6 Pipe-joints and Hough transform results

The parameter space can now be spanned by (a, b, r). For each point on the desired circle, if we draw a circle with center in this point with the desired radius (to facilitate the description, we suppose we know the radius while in fact we do not), all the circles will pass through the center of the desired circle. If we use an accumulator to count all coordinates that the perimeter of the circles we have drawn will pass through, the maximum value will be the center we desired. When we do not know the radius, it is the same, just change a 2-dimensional accumulator to a 3-dimensional one.

6.4.1 Preprocessing for Hough Transform

Although pipe-joints can present good objects for image processing, the video pictures were recorded under very difficult conditions, and there are three disadvantages inherent in these pipe-joint images:

- Low and uneven contrast around the pipe-joints due to poor illumination;
- A very dirty background owing to encrustation, debris, etc.;
- A big gap at the bottom because of residual effluent.

These features make it difficult to detect the pipe-joint boundaries. For this reason, we do the following preprocessing in our algorithm to increase the detection accuracy as well as the processing speed.

(a) Grayscale transformation
 Transform the color image to gray scale reduce the calculated amount.
(b) Edge detection
 In order to recognize the shape of the pipe-joint, we need to identify the edge of the pipe-joint as preparation for the Hough transform. In our algorithm, we use Canny edge detector because of its high efficiency.
(c) Dilatation and blurring

Because of the disadvantages inherent in the pipe-joints images, the detected edges are usually not ideal, with some gaps and noises. In order to increase the accuracy of the Hough transform, we do the dilatation and blurring to make the pipe-joint more visible and evident.

Figure 6.7 shows the result of the above processing.

The advantage of Hough transform method is that it can give an accuracy estimate of the center point although the pipe-joint is incomplete and with some noises. But for images without pipe-joints or cases where the pipe-joints are unobvious, the Hough transform will be not applicable.

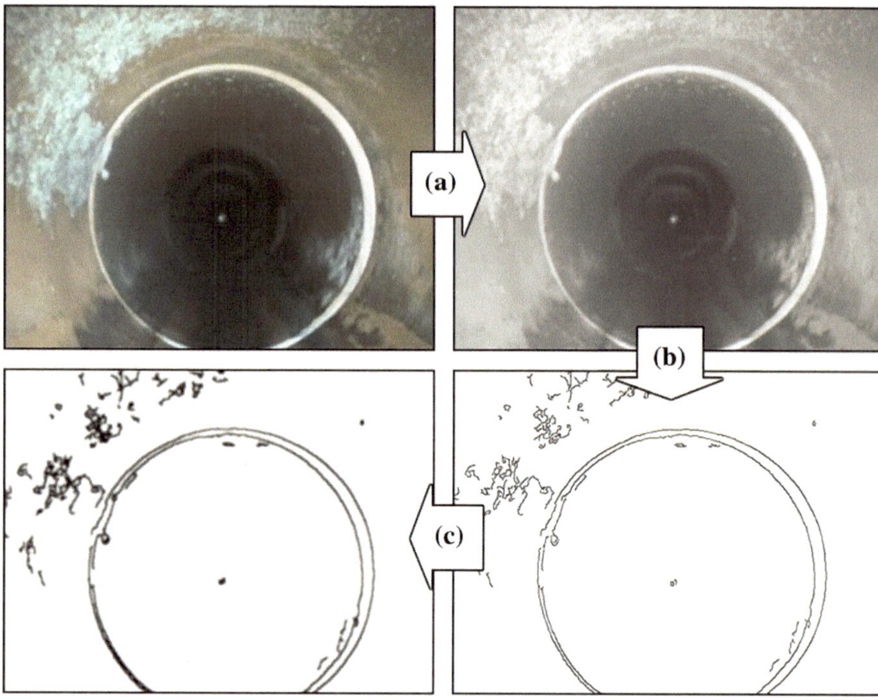

Fig. 6.7 Result of preprocessing

6.5 Least-Squares Technique

Usually, the centers of the sewer pipes are the darkest part in the pictures. The least-squares technique is the method in which we use the grayscale data to find the minimum points of the x and y coordinates by least-squares method and determine the central point. (In gray scale, all data are from 0 to 255, while 0 represents black and 255 represents white.) Figure 6.8 shows an example using the least-squares method.

Here, we use the grayscale data in the central line of each coordinate and use a second degree curve (a parabola) to approximate the given set of data. From the result, we can see the data are not fitted very well by a parabola and the estimated central point is also not very accurate. Therefore, we contemplate two improvements for this method in our algorithm.

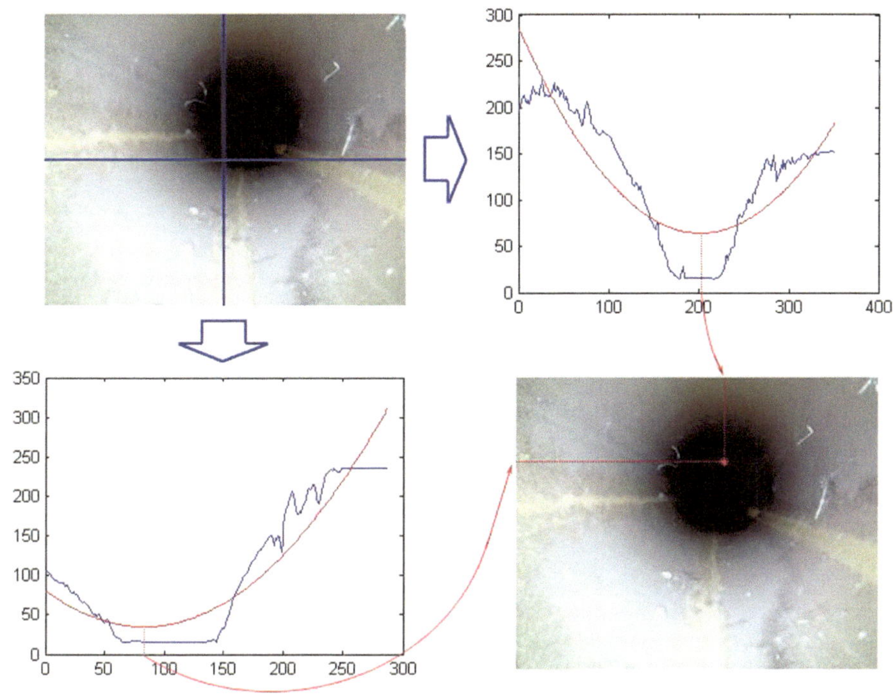

Fig. 6.8 Example of least-squares method

The first improvement is to use a higher degree polynomial to approximate the given data set. Obviously, with the increasing of the polynomial degree, the error of fitting will be reduced. But on the other hand, the computational complexity will be increased. Considering the time-consuming, we use the least squares with 4th degree polynomials.

$$y = a_0 + a_1x + a_2x^2 + \cdots + a_mx^m \tag{6.2}$$

To approximate the given set of grayscale data, refer to Fig. 6.9.

Another improvement introduced in our algorithm is using a variable scale of data set to perform the least-squares fitting. As for the influence by the lighting or the noises in the background of the image, the fitting result shows high noise. So we do the following process to cut down these noises. After finding the darkest point for one data line, we do the least-squares fitting again using the data set near this point, and then loop this process. But the more number of times we do the least-squares fitting, the more is the calculation cost. By experiment, we found two cycles of the process can already give an ideal result for the center estimation and do not cost much computation times. The result is shown in Fig. 6.10.

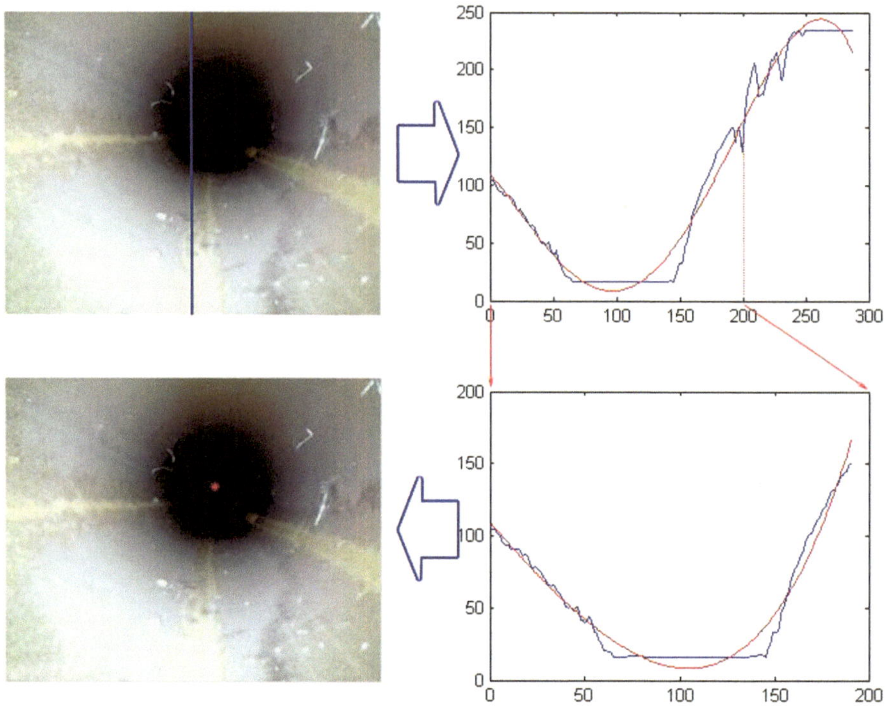

Fig. 6.9 Improved least-squares method

From the result in Fig. 6.9, we can see that the improved least-squares method give a better performance than the former one showed in Fig. 6.8.

The advantage of this method is it can be used in almost all kind of sewer pipes. However, when the sewer pipe is not very straight, the darkest point of the image may be different in comparison to the center point of the sewer pipe.

Considering the advantage and disadvantage of these two methods, we use the combination of them in our algorithm. Usually, a certain sewer pipe has its inherent characteristic whether the pipe-joint can be clearly seen or not. Just using the combination of these two methods, for one sewer pipe will cause large offset due to the different accuracy of them. So before starting the generation of the developed map, we carry out both of these two methods and determine which one is better for the specific sewer pipe. This judgment is based on the distances between the centers estimated and the center of the image. We choose the closer method as the preferred one.

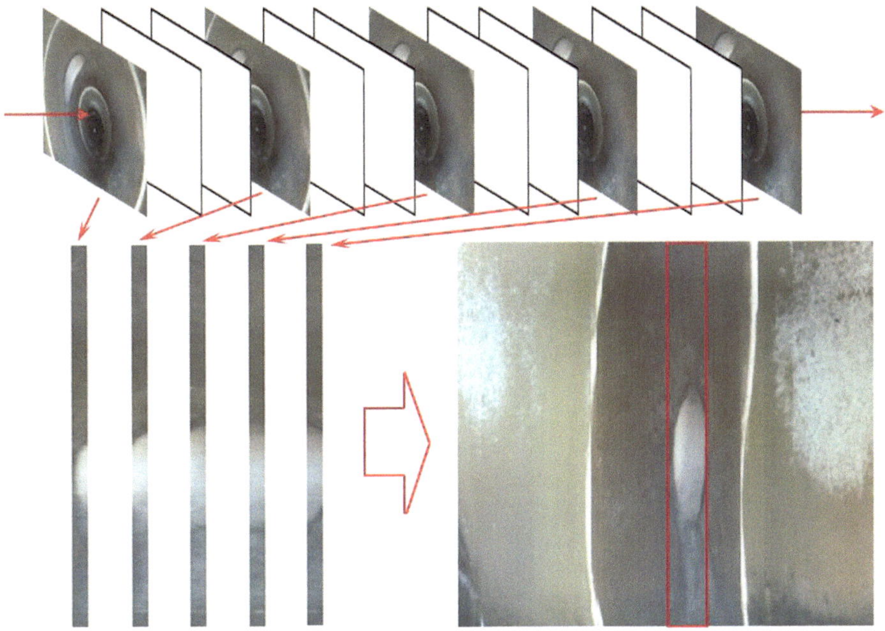

Fig. 6.10 Example of connection

6.6 Connection

Using the unfolded images of separate frames, we need to connect them to a developed map of the sewer pipe. Usually, the splicing effect is affected by the speed of the robot-inspector, the diameter of the pipe, the width of the folded band image and the number of intervals of the frames for which we do the unfolding process. It is difficult to determine a relationship of all these factors. Actually, the speed of the robot and the diameter of the sewer pipe are constant, and the intervals of the frames can be determined by the real-time constraint. (As the more frequent we grab the frame for processing, the less processing time we will have to satisfy the real-time requirement.) We just need to determine the width of the folded band to finish the connection process.

Here, by experimentation, for robot speed of 13.7 m per min, the pipe diameter of 200 mm and the intervals of every three frames (we will discuss it later), the width of the folded band image is obtained as 5 pixels. An example is given in Fig. 6.11.

So far, we have introduced the unfolding process and the connection process for the developed map. Now, let's see the complete flow of our algorithm, which is explained in Fig. 6.11.

In our algorithm, first we determine which unfolding method is preferable for the given sewer pipe. And then, for every three frames, we carry out the unfolding

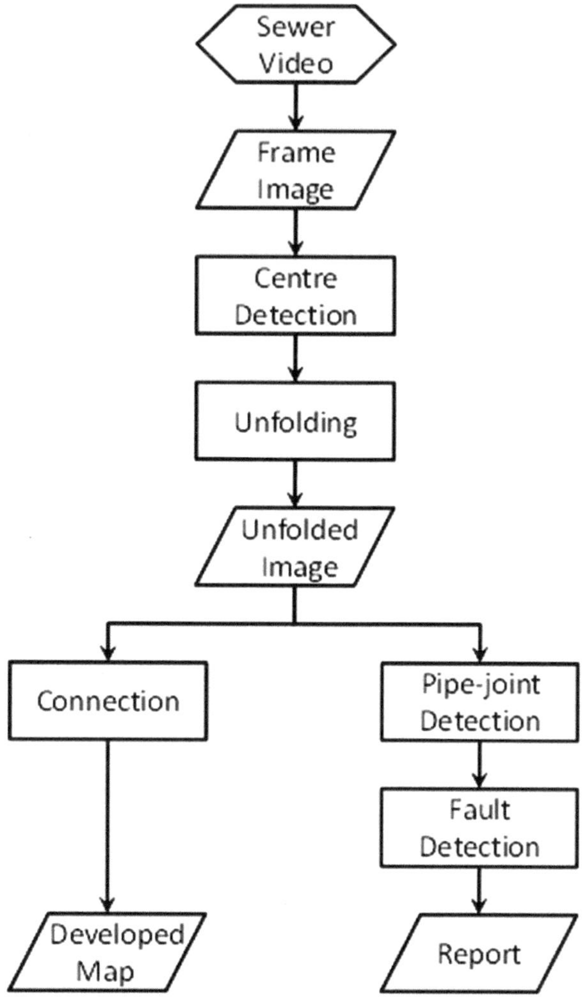

Fig. 6.11 Flowchart of the image processing algorithm

process to estimate the central point by using either Hough transform method or least-squares method and carry out a comparison with the previous one. If the deviation is to a reasonable extent, we use the new center point to replace the old one. Finally, we connect the unfolded image one by one.

Now, let's discuss the real-time constraints. Table 6.3 shows the processing times estimation for the algorithm.

For the video recording, usually it has the frame rate of 30 fps (frame per second). One frame requires 33 ms while three frames about 100 ms. As the process for one frame in our entire algorithm requires about 73.5 ms, which is less than three frame-time duration, it is possible for our algorithm to be used in real-time processing.

Table 6.3 Processing time estimation

Processing environment	
CPU	2.10 GHz × 2
Memory	1.00 GB
Language	C++
Running time (per frame) Hough/LS (ms)	
Finding center	43.8/17.1
Unfolding	28.1
Connection	1.5
All	73.5/46.7

6.7 Summary

In this chapter, we have introduced a broad algorithm for generating the developed map of sewer pipe. Toward this direction, we have first discussed about the hardware environment. Next, we have discussed about the unfolding algorithms. This was followed by discussions on image transform methods. We have finally developed a complete flow diagram for the imaging process. It may be noted that the advantages of our algorithm are mainly in two points.

First, it can identify the central point of the sewer pipe automatically and accurately, by using the combination of Hough transform and least-squares methods.

Second, it can be used in real-time detection, which makes it possible to understand the condition of the sewer pipes during the investigation.

As for further research, based on this algorithm, we plan to develop an auto visual inspection system to recognize the defects together with some auto-detection algorithms and some additional detectors for information fusion. The second issue requires development of multisensory environment which we will discuss in the subsequent chapters.

Model Questions

1. Explain the step of unfolding process of the image data.
2. Explain the coordinate transformation in Fig. 6.4.
3. Compare the interpolation algorithms using unfolded image data.
4. Explain why the Hough Transform Method use the image processing.
5. Explain the estimate method of the center point of the inside pipe image data.

References

1. Gunatilake P, Siegel MS, Jordan AG, Podnar G (1997) Image understanding algorithms for remote visual inspection of aircraft surfaces. Mach Vis Appl Ind Inspection V 2–13
2. Oosaka T, Tougou K, Numao S (2005) The development of a TV inspection system with panorama sensor for sewer pipe. Nippon Koei Tech Forum 13:77–82

3. Saito H, Date K, Mochiduki T (2003) The development of systems concerned with expansionary image of sewer pipe. Nippon Koei Tech Forum 11:01–107
4. Xu K, Luxmoore AR, Davies T (1998) Sewer pipe deformation assessment by image analysis of video surveys. Pattern Recogn 31(2):69–180
5. Kawasue K, Sakai S, Wakiyama T (2008) Sewer pipe measurement robot using parallel laser sheet. Jpn Soc Mech Eng 90–97
6. http://www.iiw.co.jp/
7. http://opencv.jp/opencv-1.0.0/document/opencvref_cv_sampling.html
8. Wu Q, Merchant FA, Castleman KR Microscope image processing, pp 53–54
9. Wang Z (2008) The design and realization of image enlargement algorithm, Bachelor Thesis. Shanghai Jiaotong University, China

Chapter 7
Introduction to Smart Materials

7.1 Introduction

Structural health monitoring using smart sensors and actuators largely relies on the appropriate choice of smart materials. The smart materials can be categorized into four major groups based on the driving mechanisms; these are piezoelectric, magnetostrictive, phase-transition dependent and electro/magnetorheological materials [1–5]. The existence of cross-coupling between mechanical fields (like stress/strain/viscosity) and other fields (like electric/magnetic/thermal) makes these materials very useful for sensing, actuation, and control. In Table 7.1, a group of five input–output pairs are shown along with the cross-coupling in material properties, which are commonly exploited for smart sensing and actuation.

Piezoelectric materials are initially used as transducer for sensing dynamic changes in wave pressure in order to detect submarines. These materials develop electric charges when deformed due to mechanical stress (known as direct piezoelectric effect) and conversely get deformed by the application of electric potential. Figure 7.1 shows the deformation in a piezoelectric body due to the application of voltage 'V' in an exaggerated scale. In this figure, the characteristic lengths of the body are denoted by l (length), w (width), and t (thickness). Similar coupling effects of mechanical and magnetic field are found in magnetostrictive materials. Thus, these two groups of smart materials could be used both as actuators (to generate mechanical deformation through magnetic/electric signal) and as sensors (to generate electromagnetic field if mechanically deformed). Phase-transition-dependent materials and smart rheological fluids, on the other hand, are known generally for actuation purposes only. The shape memory alloys with phase-transition-dependent property show large deformation during a solid–solid phase-transition process from martensite to the parent austenite state [4]. Interestingly, the phase-transition depends on both stress and temperature such that by changing either of them, volume change can be effected. The electrorheological fluids (ERF) are specially formulated suspensions. It shows a large change in the resistance to flow (viscosity)

H. Ogai and B. Bhattacharya, *Pipe Inspection Robots for Structural Health and Condition Monitoring*, Intelligent Systems, Control and Automation: Science and Engineering 89, https://doi.org/10.1007/978-81-322-3751-8_7

Table 7.1 Sources of material intelligence

Input \ Output	Current/Charge	Magnetization	Strain	Temperature	Light
Electric Field	Conductivity Permittivity	Electromagnetic Effect	Reverse Piezoelectric Effect SA	Ohmic Resistance	Electro-Optic effect
Magnetic Field	Eddy Current Effect	Permeability	Joule Effect Magnetostriction SA	Magneto-caloric Effect	Magneto-Optic effect
Stress	Direct Piezoelectric Effect SS	Villary Effect SS	Elastic Modulus	Thermo-Mechanical Effect SS	Photo-elastic Effect SS
Heat	Pyroelectric Effect	Thermo-magnetization	Thermal Expansion/Phase-transition SA	Specific Heat	Thermo-luminescence
Light	Photo-voltaic Effect	Photo-magnetization	Photostriction SA	Photo-thermal effect	Refractive Index

Note SS denotes smart sensor, and SA denotes smart actuator

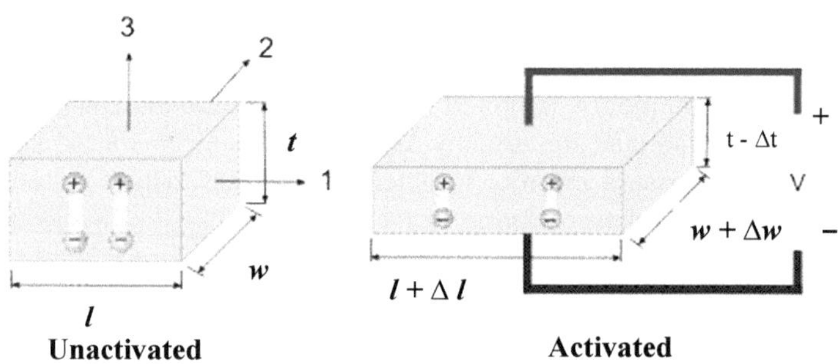

Fig. 7.1 Deformation of a piezoelectric body due to applied voltage

due to an applied electric field [5]. Similar characteristics are shown by suspensions of ferromagnetic particles in silicone fluid on application of magnetic field; these are called magnetorheological fluids (MRF).

The nature of response while subjected to mechanical/electrical/magnetic field-based excitation makes a smart material suitable for a specific field of application. Shape memory alloy, for example, is slower in response, but the response magnitude is very high. Hence, these are more suitable for static shape control and low-frequency dynamic excitation, whereas magnetostrictive and piezoelectric materials are more useful as actuator or sensor in medium to large bandwidth vibration control applications. Again, electro- and magnetorheological fluids have

Table 7.2 Salient properties for smart material

Properties important for actuation	Piezoelectric material		Magnetostrictive material		
			Phase-transition-dependent material		
Piezo-ceramic	PVDF		Terfenol-D	Nitinol	FSMA
Maximum free strain (Λ) in microns	2000	700	2000	20,000	30,000
Young's modulus (GPa)	60–70	2–3	48	27.5 M-phase, 90 A-phase	0.45–0.82
Bandwidth	0.1 Hz–GHz	0.1 Hz–GHz	0.1 Hz–10 kHz	0–10 Hz	100 Hz

found many applications in vibration isolation. A comparison of the actuation potential of some typical smart materials is given in Table 7.2 [6–8]. It is interesting to note that although shape memory alloy (SMA), like nitinol, produces high strain, the time constant is too large to be used as active damper for vibration control applications [9]. A new variation of SMA known as ferromagnetic SMA (FSMA) shows better dynamic performance in comparison with SMA. For FSMA, the phase-transition is initiated by magnetic field.

Piezo-ceramic plates and magnetostrictive Terfenol-D rods are more attractive due to their good operational bandwidth and moderate capacity of generating actuation strain. Polyvinylidene fluoride (PVDF), yet another type of piezoelectric polymeric material, is relatively inferior as an actuator, since it has poor structural strength and the strain induction is also quite small. However, PVDF can be used as a very good sensor, as it interferes minimally with the host structural response. Often, optical fiber and magnetorheological fluid are also considered in the domain of smart materials [10, 11] among which the optical fiber is notable for sensing applications only since it possesses one-way coupling between mechanical and optical fields.

7.2 Piezoelectric Material

Piezoelectric effect, signifying generation of 'electricity from pressure' was discovered by Pierre and Jacque Curie more than 100 years ago. The so-called direct effect of charge generation in a class of crystals while subjected to mechanical stress was first observed in the crystals such as zinc blend, sodium chlorate, tourmaline, and quartz [12]. The 'inverse effect' in piezoelectricity, signifying the deformation in crystals on the application of a voltage, was theoretically predicted by Lippman showing it as a consequence of a reversible thermodynamics process and later experimentally validated by Jacque Curie. Voigt did the first rigorous analysis of correlating crystal structure symmetry to piezoelectricity. It was shown that only 11 crystal groups could have piezoelectricity out of the possible 32 crystal groups (based on simple lattice geometry and symmetry operation on crystal faces). This property is retained by the crystals up to a critical temperature known as Curie temperature. The uniqueness of these groups of piezo-crystals is their non-centrosymmetric structure. A subgroup of such crystals always falls to a state of spontaneous polarization. The phenomenon widely known as 'ferroelectricity' occurs due to the presence of hydrogen bond dipoles in the elements. Over the years, guided by these characteristics, material scientists have successfully identified many crystals with piezoelectric property. Most of them are found unfit for engineering applications due to their high-temperature sensitivity and large hysteresis loop. It is only with the invention of Barium Titanate ceramic (around 1947) that building of efficient solid-state transducer has been made possible. Soon afterward, Barium Titanate is substituted by Lead Titanate Zirconate (commonly known as PZT) as it has higher Curie temperature and possesses stronger piezoelectric effect [13]. Similar effect has also been discovered in a polymeric crystal called Polyvinylidene fluoride (PVDF) by Kawai [14]. Most of the developments in smart structural technology are centered on these two piezoelectric materials.

The fundamental equations of piezoelectricity are given by Eqs. (7.2.1) and (7.2.2) respectively

$$\sigma_{ij} = C^E_{ijkl} S_{kl} - e_{kij} E_k \tag{7.2.1}$$

$$D_i = e_{ikl} S_{kl} + \varepsilon^S_{i\,j} E_j \tag{7.2.2}$$

where the subscripts i, j, k, l = 1, 2, 3 are tensor indices. The stress tensor is represented here as σ, S—the strain tensor, E—the electric field intensity, and D—the electric field displacement. The integration of D over the electrode area provides the total charge accumulated on the electrode. C^E is the elastic stiffness matrix where the superscript E denotes the elastic constant measured under constant electric field (henceforth, the superscript will not be shown explicitly for the sake of brevity); e—the piezoelectric stress–charge matrix; and ϵ—the permittivity matrix; similar to C, it is measured under constant strain condition. The details of the axes are shown in Fig. 7.2.

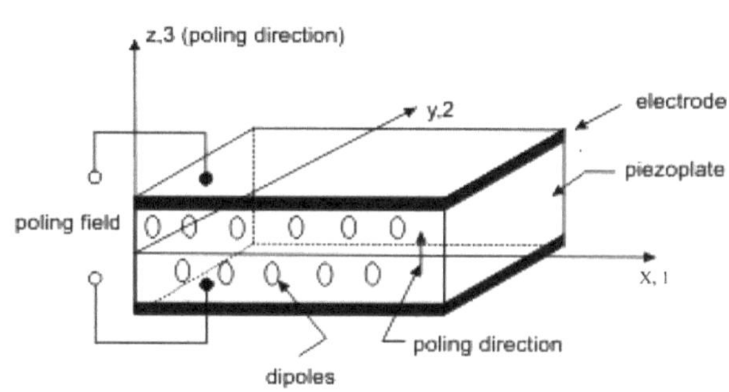

Fig. 7.2 Poling of piezoelectric material

Commonly, piezoelectric materials show dihexagonal symmetry in crystal structure. Hence, the material symmetry conditions as shown in Eq. (7.2.3) could be applied to the constitutive relationship

$$
\begin{aligned}
C_{ijkl} &= C_{jikl} = C_{klij} \\
e_{kij} &= e_{kji} \\
\varepsilon_{ij} &= \varepsilon_{ji}.
\end{aligned}
\tag{7.2.3}
$$

After experimentally obtaining the material constants and using the material symmetry for hexagonal crystal structure, Eqs. (7.2.1) and (7.2.2) could be simplified and written as:

$$
\begin{Bmatrix} \sigma_x \\ \sigma_y \\ \sigma_z \\ \sigma_{yz} \\ \sigma_{xz} \\ \sigma_{xy} \\ D_1 \\ D_2 \\ D_3 \end{Bmatrix} =
\begin{bmatrix}
C_{11} & C_{12} & C_{13} & 0 & 0 & 0 & 0 & 0 & -e_{31} \\
C_{12} & C_{11} & C_{13} & 0 & 0 & 0 & 0 & 0 & -e_{31} \\
C_{13} & C_{13} & C_{33} & 0 & 0 & 0 & 0 & 0 & -e_{33} \\
0 & 0 & 0 & C_{44} & 0 & 0 & 0 & -e_{15} & 0 \\
0 & 0 & 0 & 0 & C_{44} & 0 & -e_{15} & 0 & 0 \\
0 & 0 & 0 & 0 & 0 & C_{66} & 0 & 0 & 0 \\
0 & 0 & 0 & 0 & e_{15} & 0 & \varepsilon_1 & 0 & 0 \\
0 & 0 & 0 & e_{15} & 0 & 0 & 0 & \varepsilon_2 & 0 \\
e_{31} & e_{31} & e_{33} & 0 & 0 & 0 & 0 & 0 & \varepsilon_3
\end{bmatrix}
\begin{Bmatrix} S_x \\ S_y \\ S_z \\ S_{yz} \\ S_{xz} \\ S_{xy} \\ E_1 \\ E_2 \\ E_3 \end{Bmatrix}
\tag{7.2.4}
$$

The electromechanical coupling is outlined inside the boxes in Eq. (7.2.4). It may be noted that axes *x*, *y*, and *z* corresponding to the mechanical parameters are identical with 1, 2, and 3 used for the electrical system.

Consider a thin slab of piezoelectric element shown in Fig. 7.2. In the presence of a high electric field (poling field), the randomly oriented dipoles get aligned to the direction of the electric field. Consequently, the strain induced is magnified. The direction of alignment is known as the poling direction. For a thin plate, we may ignore the normal stress σ_z and the shear stresses σ_{xz} and σ_{yz} too may be neglected for plane stress assumption. Hence, Eq. (7.2.4) could be further simplified as:

$$
\begin{Bmatrix} S_x \\ S_y \\ S_{xy} \\ D_3 \end{Bmatrix} =
\begin{bmatrix}
1/E_p & -\upsilon/E_p & 0 & -d_{31} \\
-\upsilon/E_p & 1/E_p & 0 & -d_{32} \\
0 & 0 & 2(1+\upsilon)/E_p & 0 \\
d_{31} & d_{32} & 0 & \varepsilon_{33}
\end{bmatrix}
\begin{Bmatrix} \sigma_x \\ \sigma_y \\ \sigma_{xy} \\ E_3 \end{Bmatrix}
\tag{7.2.5}
$$

E_p is the modulus of elasticity of the piezoelectric material, ν is the Poisson's ratio, and d_{ij} are the piezoelectric strain–charge constants. The first three equations of (7.2.5) are generally used as the constitutive equations of piezoelectric actuators, while the last equation is used to model piezoelectric sensors.

From Eq. (7.2.5), it is clear that if a piezoelectric thin slab is subjected to mechanical load, the total strain S developed in an active layer would consist of two parts—the structural or elastic strain S_s and the piezoelectric strain S_a such that

$$S = S_s + S_a \tag{7.2.6}$$

where $S_a = [-d_{31}E_3, -d_{32}E_3, 0]^{\mathrm{T}}$.

The structural strain on the other hand will depend on the nature of loading and boundary condition applied to the piezo-plate.

It is also evident from Eq. (7.2.5) that for poling across the lamina (see also Fig. 7.1), the active strain is developed only in the normal plane. To generate axial strains in the thickness direction, ceramics with different crystal cuts are used. These are commonly known as piezo-stacks. The electroelastic coupling components in the 3-3 directions, like d_{33} or e_{33}, become important in such cases. Another important parameter used in piezoelectricity is the voltage constant g (defined in Sect. 7.3). The higher value of g signifying higher voltage sensitivity implies more suitability of the material for sensing application. The piezoelectric constants and their significance in active vibration control are discussed in the following section.

Example 7.2.1 A thin piezoelectric plate of size 25 mm × 10 mm and thickness 1.5 mm has electroding on top and bottom as shown in Fig. 7.3 (with an exaggerated view of the thickness). Assume the piezoelectric material to have elastic modulus of 65 GPa, Poisson's ratio 0.3, and electromechanical coupling coefficients, d_{31} and d_{32}, to be -50×10^{-12} and 200×10^{-12} m/V, respectively. Find out the strains in the plate when it is subjected to a voltage of 300 V and a force of 20 N.

Solution

Following Fig. 7.3 and Eq. (7.2.6), the active strains in the plate are as follows:

$$S_{a_x} = -\frac{d_{31}V}{t} = \frac{50 \times 10^{-12}\,(\mathrm{m/V}) \times 300\,(\mathrm{V})}{1.5 \times 10^{-3}\,(\mathrm{m})} = 1 \times 10^{-5} = 10\,\mu\text{-strain}$$

since $d_{31} = d_{32}$, $S_{a_y} = S_{a_x} = 10\,\mu$-strain. Also, since a tensile force of 20 N is acting on the plate along the x-direction, the structural strain along x-direction is $S_{s_x} = \frac{20\,(\mathrm{N})}{15\times10^{-6}\,(\mathrm{m}^2)\times65e^{12}\,(\mathrm{N/m}^2)} = 0.02\,\mu$-strain, and the structural strain along

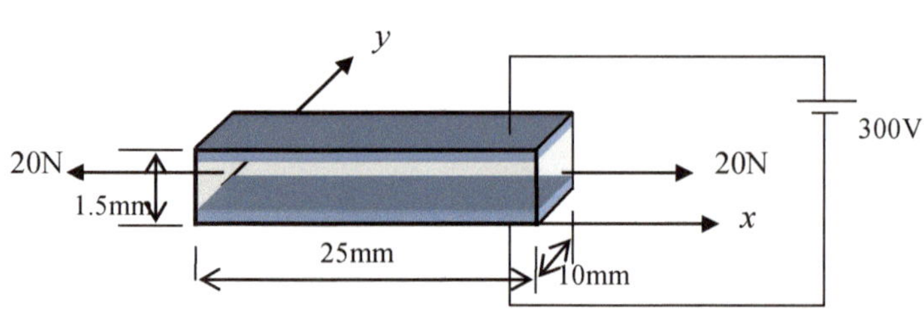

Fig. 7.3 Actuation of a piezo-plate

y-direction is $S_{s_y} = -0.3 \times S_{s_x} = -0.006$ μ-strain. Hence, the total strain along x-direction is 10.02 μ-strain, and the total strain along y-direction is 9.99 μ-strain.

7.3 Piezoelectric Coefficients

Four constants are frequently used for the comparison of the performances of different piezoelectric materials for sensing and actuation. These are piezoelectric charge constant (d), piezoelectric voltage constant (g), electromechanical coupling factor (k), and the frequency constant (N_p).

The piezoelectric charge constant, d, expressed in 'm/V' or 'pC/N' (1 pico-Coulomb (pC) = 10^{-12} Coulomb), is defined by the following simple relationship:

$$d_{31} = \frac{\Delta l/l}{V_3/t} = \frac{q}{F_1}, \quad d_{32} = \frac{\Delta w/w}{V_3/t} = \frac{q}{F_2}, \quad d_{33} = \frac{\Delta t/t}{V_3/t} = \frac{q}{F_3} \tag{7.3.1}$$

where the geometric parameters are explained in Fig. 7.1. Here, 'q' denotes the charge collected in the electrode surfaces, F_i, $i = 1...3$, denote the forces along the respective directions, and V_3 denotes the voltage applied along the z-direction. The first expression of coupling coefficients signifies d as the mechanical strain experienced by a piezoelectric element for an applied voltage V, while the second relationship defines d as the induced polarization generated by a force F. The piezoelectric constants d_{31}/d_{32} are usually negative signifying the generation of positive strain upon the application of positive voltage.

The piezoelectric voltage constant, g, expressed in V-m/N, is similarly defined as:

$$g_{31} = \frac{V_3}{F_1/w}, \quad g_{32} = \frac{V_3}{F_2/l}, \quad g_{33} = \frac{V_3/t}{F_3/(w \times l)} \tag{7.3.2}$$

where V_3 is the voltage sensed along the z-direction due to the application of pressure. Like d_{31}, g_{31} is also usually negative signifying the generation of positive voltage upon the application of tensile force to the system.

In general, a relatively large value of d indicates the superiority of the piezoelectric material as an actuator, and similarly, a higher value of g indicates better suitability of the material as a sensor. However, neither of these coefficients describes the quality of a piezoelectric material as an efficient energy converter. For example, hard PZT has higher d_{31} than soft PZT, whereas the soft PZT has higher efficiency in energy conversion. The electromechanical coupling factor, k, measures the electromechanical energy conversion efficiency. It is expressed by the simple relationship

$$k^2 = \frac{d_{33}^2}{S^E \varepsilon^T}. \tag{7.3.3}$$

The superscripts, E and T, signify the values of S and ε under constant electric field and constant stress condition, respectively.

When an unconstrained piezoelectric material is subjected to an alternating current, the material shows resonating behavior at certain frequencies. For a disk, piezoelectric element of diameter D_p, and thickness t, the resonating frequencies corresponding to the radial and axial modes (f_r and f_a, respectively) are expressed in terms of the two frequency constants N_d and N_t as

$$N_d = f_r D_p, \quad N_t = f_t t. \tag{7.3.4}$$

The frequency constants are important for applications of piezoelectric material as a mechanical resonator. Typically, such resonances occur at driving frequencies beyond 20 MHz and hence are of less importance in vibration control. High-frequency applications of piezoelectric resonators include piezoelectric ultrasonic motors and piezoelectric probes. The other three coefficients (d, g, and k), however, play crucial roles in vibration control applications. Electromechanical properties of different piezoelectric materials are listed in Table 7.3. Here, Λ represents the free strain in a piezoelectric actuator without any constraining force.

Table 7.3 reveals that the composite of PZT–PVDF has high electromechanical coupling with a moderate density which is in between those of PZT and PVDF. The elastic modulus is also seen to be quite high in comparison with that of PVDF. Thus, such composites present a good trade-off between excellent actuation potential of PZT and sensing capability of PVDF. Often, the high-actuation strain generating capability of SMA is exploited by doping elements of SMA into piezo-ceramic PLZST. The product is known as shape memory ceramic active material. As high as 6000 μ-strain along with memory effect is achieved through this material. However, such materials are still in the developmental stage, and hence, their commercial/engineering viability is yet to be established.

Table 7.3 Salient properties of various piezoelectric materials

Property	PZT (hard)	PZT (soft)	PZT-PVDF	PMN-PT	$LiNbO_3$	PVDF
d_{33} (pC/N)	190	425	120	1240	6	30
d_{31} (pC/N)	−55	−170	–	–	−0.85	−16
g_{33} (mV m/N)	54	27	300	43		150
g_{31} (mV m/N)	−16	−11	–	–		−150
k_{33}	0.67	0.70	0.80	0.92	0.17	0.11
E_p (GPa)	63	45	~30	100	20	2.7
Density (ρ) (kg/m^3)	7500	7500	3300	8120	4600	1760
Λ	1500	1980	400	3100	1210	700

7.4 Piezoelectric Sensors and Actuators

PZT-based actuators can normally generate a maximum strain of about 0.2% (about 2000 μ-strain). Though single crystals of PZN and PMN are known to generate strains of the range of 8000 μ-strain, the use of such crystals as actuators is limited due to their high cost and difficulty of integrating in a structure. Even with the maximum available active strain, it is quite difficult to control machine vibration using the smart materials directly. Various forms of displacement and force amplification techniques have been recently developed toward the use of piezoelectric actuators in industrial scale. These systems are classified into three major groups [15]:

(i) Internally Leveraged System: In this system, the actuators contain multiple piezoelectric elements to get an amplified effect. The simplest example is a piezo-stack where many piezoelectric wafers are stacked in such a way that a comparatively larger deformation is obtained in the d_{33} mode by applying a smaller voltage. More advanced systems include various configurations such as Rainbow, C-block, and Crescent as detailed in the Appendix.

(ii) Externally Leveraged System: In these actuators, mechanical systems are utilized to amplify the output of a piezoelectric actuator—these include actuators such as unimorph, bimorph, flexure-based actuator, moonie, cymbal .

(iii) Frequency Leveraged System: This type of system is based on alternating current supply to a piezo-actuator. Typical examples are piezoelectric inchworm motors, ultrasonic motors.

Even though different designs of actuators achieve different degrees of efficiencies in terms of energy conversion, displacement, and force amplification, the basic principle used for strain generation is quite similar. Several models have been developed for this purpose, and two of them are explained here.

7.4.1 *Uniform Strain Model of Induced Strain Actuation*

The static models developed by Crawley and Anderson [16] are based on two different assumptions on distribution of strain across the thickness of a typical beam. In the 'uniform strain' model, it is assumed that the strain remains constant across the piezo-actuator, while it varies linearly inside the substructure. The model has been used for surface-bonded actuation. In 'Bernoulli–Euler' model (see Sect. 7.4.2), on the other hand, a linear variation of strain is assumed for the entire cross section which is considered for embedded actuation. For each of these models, the actuators embedded/bonded on top and bottom of the beam are excitable in the same phase to cause uniform extension or contraction. Otherwise, bending can also be generated through out-of-phase excitation of the piezo-ceramics. Figure 7.4a–d shows detailed

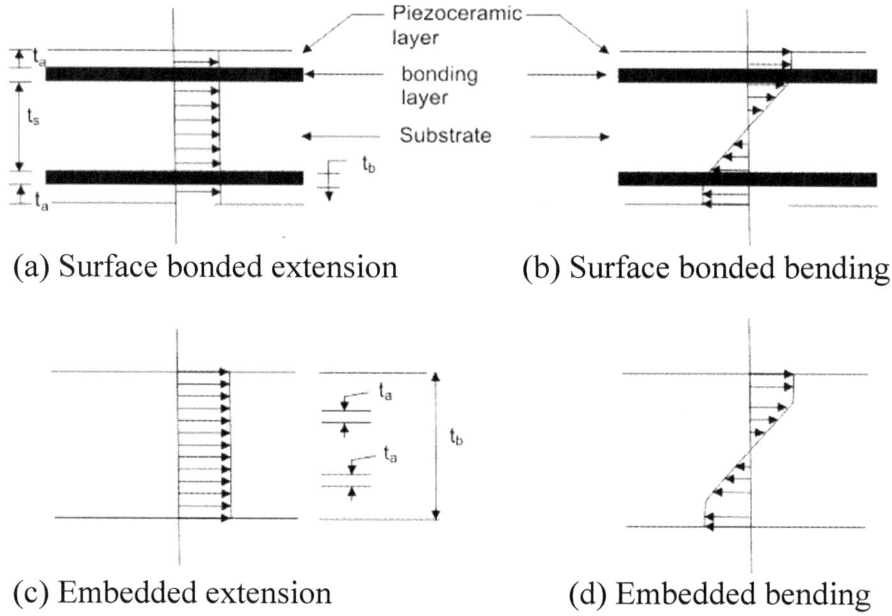

Fig. 7.4 Stress distribution across the beam thickness corresponding to various placement of the piezo-layer

sketches of all the strain diagrams. For surface-bonded actuation, the static equilibrium corresponding to the induced strain can be expressed as

$$2F_a + F_s = F_i \tag{7.4.1}$$

where F_a is the reactive force developed in each active layer, F_s is that in the substrate, and F_i is the total force. From Eq. (7.4.1), one obtains

$$2\left(E_p S_a A_p\right) + E_s S_s A_s = 2\Lambda E_p A_p \tag{7.4.2}$$

where S denotes the strain, E the modulus of elasticity, A the area of cross section, and Λ the free strain. The subscript p denotes the piezoelectric material, and s denotes the substrate. Since it is assumed that near the actuator–substrate interface, strain remains unchanged, we can write

$$S_a = S_s = \frac{2\Lambda}{2 + \Psi_e} \tag{7.4.3}$$

where the in-plane stiffness ratio is $\psi_e = (E_s\ A_s)/(E_p\ \ A_p)$. Similarly, for bending, considering equilibrium of active and reactive moments, one gets

$$E_p\, S_a\, A_p t_s \;+\; \frac{2E_s I_s S_a}{t_s} \;=\; \Lambda E_p\, A_{\,p} t_s \tag{7.4.4}$$

Once again the strain compatibility at the interface leads to

$$S_a \;=\; S_s \;=\; \frac{6\Lambda}{6+\psi_b} \tag{7.4.5}$$

with the bending stiffness ratio ψ_b = 12 $(EI)_s$ / [t_s^2 $(EA)_p$]. However, when the thickness of the bonding layer is finite, the presence of viscoelastic bonding material reduces the transmission of stress from actuator to substrate and the induced strain to actuation strain ratio is shown to be given by the following relationships [16]

$$S_a \;=\; \frac{\alpha}{\alpha+\psi}\Lambda\left[1+\frac{\psi\,\cosh(\Gamma\bar{x})}{\alpha\,\cosh(\Gamma)}\right] \tag{7.4.6}$$

$$S_s \;=\; \frac{\alpha}{\alpha+\psi}\Lambda\left[1-\frac{\cosh(\Gamma\bar{x})}{\alpha\,\cosh(\Gamma)}\right] \tag{7.4.7}$$

with

$$\Gamma^2 \;=\; \overline{D}\,\frac{\alpha+\psi}{\psi}\,, \qquad \overline{D} \;=\; \frac{(G/E_a)\,(t_b/t_p)}{(t_b/l_p)^2} \tag{7.4.8}$$

and $\bar{x}$ is the non-dimensional length parameter varying from −1 to +1 (edge to edge of the actuator). The geometric constant α is 2 for extension and 6 for bending, and ψ is the stiffness ratio related to bending or extension based on the appropriate case. G is the shear modulus of the bonding layer, and Γ is the shear lag contributed by the bonding layer of thickness t_b. A high value of Γ signifies a thin layer with stiff bonding. As $\Gamma \to \infty$, the above equations become identical to the earlier set of Eqs. (7.4.3) and (7.4.5) indicating perfect bonding. Using Eqs. (7.4.6)–(7.4.8), one can also obtain the shear force for perfect bonding, generated by a pair of smart patches fixed at the top and bottom surface of a host beam of width b as

$$F_a \;=\; \frac{E_b\, t_b\, b}{\psi+\alpha}\,\Lambda \tag{7.4.9}$$

Example 7.4.1 In a bimorph configuration with smart actuators, a host specimen has two actuators placed at the top and bottom of the specimen. One can apply opposite voltages into these actuators and bend the beam either way. A fixed-free bimorph beam suitable for MEMS (micro-electromechanical system) application consists of a host beam of length 2 mm, width 200 μm, and thickness 100 μm. Two 50-μm-thick piezoelectric actuators of same length and width as the host beam are fixed on the top and bottom of the beam. The piezoelectric material has elastic modulus of 65 GPa and the electromechanical coupling coefficient, $d_{31} = -50\ 10^{-12}$ m/V. The host beam is made of silicon and of elastic modulus 100 GPa. The top actuator is actuated with a voltage of 100 V, and the bottom one is reversely connected to a −100 V source. Find out the deflection at the free end of the bimorph beam.

Solution

The second moment of area of the host beam is $I_s = \frac{200\times10^{-6}\times(100\times10^{-6})^3}{12} = 16.7\times10^{-18}\ \mathrm{m}^4$, and the flexural rigidity is $(EI)_s = 16.67 \times 10^{-6}\ \mathrm{Nm}^2$. The bending stiffness ratio of the host beam and the piezoelectric actuator is expressed as $\psi_b = 12\ (EI)_s\ /\ [t_s^2\ (EA)_p] = 1.54$. Also, for bending, $\alpha = 6$. The free strain Λ is $d_{31}V/t_a = 100\ \mu$-strain.

Using Eq. (7.4.9), the force acting on the top of the host beam at its free end is

$$F_a = \frac{100\times10^9\times100\times10^{-6}\times200\times10^{-6}}{(6+1.54)}\times100\times10^{-6} = 0.026\,\mathrm{N}$$

Moment applied at the free end of the cantilever beam is $F_a t_b = 2.652 \times 10^{-6}$ Nm. Hence, the deflection at the tip is $(ML^2/2EI) = \frac{2.652\times10^{-6}\times(2\times10^{-3})^2}{2\times100\times10^9\times16.67\times10^{-18}} = 3.18\ \mu\mathrm{m}$

7.4.2 Euler–Bernoulli Model of Induced Strain Actuation

Another modeling of smart beam for static analysis of induced strain, specifically used for embedded smart patches, is based on the assumption of linear strain variation along the entire cross section of the beam (Euler–Bernoulli model). Following Euler–Bernoulli assumptions, the strain at any point at a distance z from the neutral axis of the structure can be expressed as

$$S_t = S_s - z\kappa, \tag{7.4.9}$$

where S_s is the mid-plane strain, and κ is the curvature of the beam. Also, the total stress at any point consists of the combination of elastic stress and induced stress and is given by

$$\sigma_t = \sigma_s + E_a\Lambda, \tag{7.4.10}$$

where the subscripts s and a stand for the passive and active substrates, respectively. Using the last two equations for strains and stress and integrating over the cross section of the smart beam, the equations of force and moment balance corresponding to perfect bonding may be obtained as

$$\begin{bmatrix} EA & EB \\ EB & EC \end{bmatrix} \begin{Bmatrix} S_s \\ \kappa \end{Bmatrix} = \begin{Bmatrix} P_s + P_\Lambda \\ M_s + M_\Lambda \end{Bmatrix}, \tag{7.4.11}$$

where

$$(EA) = \int {}_zE(z)\,\mathrm{d}z, \quad (EB) = \int {}_zE(z)z\,\mathrm{d}z,$$

$$(EC) = \int {}_zE(z)z^2\,\mathrm{d}z, \quad P_s = \int {}_z\sigma(z)\mathrm{d}z, \quad M_s = \int {}_z\sigma(z)z\mathrm{d}z,$$

$$P_\Lambda = \int {}_zE_a(z)\Lambda(z)\mathrm{d}z \quad \text{and} \quad M_\Lambda = \int {}_zE_a(z)\Lambda(z)z\mathrm{d}z.$$

Solution of Eq. (7.4.11) for zero applied load condition yields the same relationship between S_s and Λ as in Eq. (7.4.3). However, the parameter Ψ_e now becomes

$$\Psi_e = \frac{(E_S A_S - E_S A_a)}{E_a A_a}. \tag{7.4.12}$$

It is to be noted that the substitution of passive material by the active piezo-layer has been taken into account in this model. In the case of bending too, the solution of Eq. (7.4.11), corresponding to zero applied loading, provides actuation strain in terms of free active strain Λ in a relation similar to Eq. (7.4.5).

7.5 Piezoelectric Sensors

Piezoelectric sensors are usually of two types—dedicated piezoelectric sensors and self-sensing actuators. The dedicated sensors follow the relationship between electric displacement and mechanical stress as described by the last one of equation set (7.2.5). The generated charge could be obtained as

$$q = \int_A D_3\,\mathrm{d}A,$$

where A is the area covered by the sensor electrode. For one-dimensional sensor, the relationship between the sensed voltage and strain can be written as:

$$V_s = \frac{d_{31}\,E_p\,A\,S_1}{C_p}, \tag{7.5.1}$$

where C_p is the capacitance of the sensor.

Similar to electromagnetic actuation, in the case of piezoelectric materials, the actuator could be directly used as a sensor. One of the earliest implementations of the self-sensing technique was in voice coils to reduce distortion and increase damping in the resonance modes of the speakers. More recently, Dosch et al. [17] have demonstrated the applicability of actuators as sensors in smart structures. The technique of self-sensing is based on the unique property of piezoelectric crystals which can undergo a concurrent sensing of strain along with generating stresses following the control signal. The property can be explained simply through the one-dimensional constitutive law of piezoelectricity as follows:

$$\begin{Bmatrix} S_1 \\ D_3 \end{Bmatrix} = \begin{bmatrix} 1/E_p & d_{31} \\ d_{31} & \varepsilon_{33} \end{bmatrix}_3 \begin{Bmatrix} \sigma_1 \\ E_3 \end{Bmatrix}, \tag{7.5.2}$$

where the terms have already been explained in Sect. 7.2. Given the values of the constants d_{31}, E_p, and ε_{33}, from the measurement of D_3 and E_3, i.e., the free charge on the piezoelectric plate and the applied voltage, S_1 and σ_1 can be computed using the aforementioned equations.

A simple electrical network through which such concurrent sensing can be achieved is shown in Fig. 7.5. The piezoelectric material is modeled as a voltage generator V_p in series with a capacitor C_p. V_p takes into account the piezoelectric properties of the material, whereas C_p accounts for the dielectric properties of the material. V_c is the voltage applied by the controller to the piezoelectric actuator. The bridge circuit eliminates the effects due to this control signal V_c by adding in

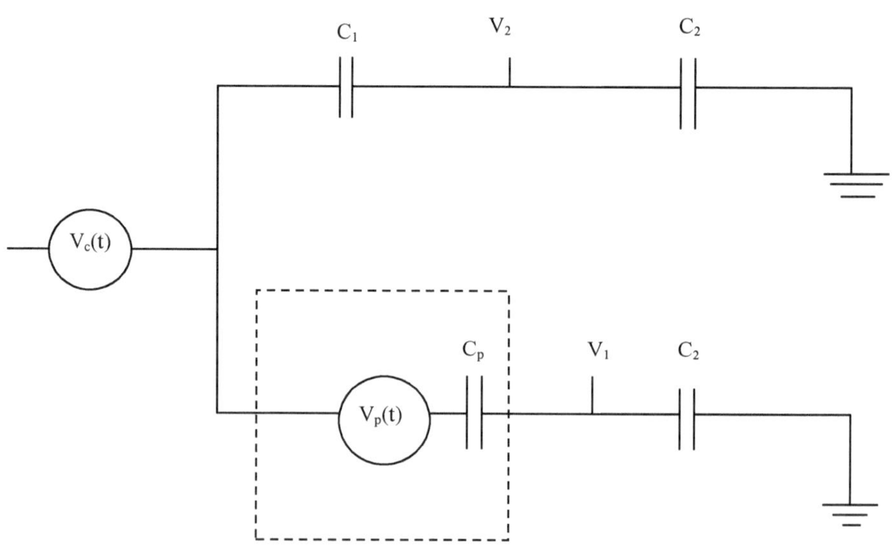

Fig. 7.5 A simple circuit for self-sensing of piezo-transducer

parallel a capacitor C_1 having equal capacitance to that of C_p. The net signal achieved by subtracting V_2 from V_1 represents the signal corresponding to the strain at the point. Thus,

$$V_s(t) = V_1(t) - V_2(t) = \frac{C_p}{C_p + C_2} V_p(t) \tag{7.5.3}$$

Vipperman and Clark [18] have implemented an adaptive piezoelectric sensor actuator based on this principle. For a broadband stochastic input, the adaptive algorithm can be used to measure the dynamic capacitance of the transducer and can be fed back to the closed loop system.

7.6 Magnetostrictive Material

Magnetostriction is mostly found in the magnetic transition materials such as iron, cobalt, and nickel and also in the rare earth materials such as lanthanum and terbium. The grains of these materials consist of numerous small randomly oriented magnetic domains, which can rotate and align under the influence of an external magnetic field. The magnetic orientation or alignment brings forth internal strain in the material, which is known as magnetostriction. Similar effect is also found in the dielectrics under high electric field, which is known as electrostriction. These electromechanical phenomena are quite different from the piezoelectricity as these are essentially nonlinear in nature, and under unbiased field, the response is always unidirectional. In other words, the materials can only expand irrespective of the direction of the magnetic field applied to it.

The phenomenon of magnetostriction was discovered in nickel by James Joule in 1840. It was also observed later in other ferromagnets and their alloys, although the maximum achievable strain was limited to 150 μ-strain only. Soon after, the discovery of low-temperature magnetoelasticity in rare earth elements, such as Tb (terbium), Dy (dysprosium), and Sm (samarium), has given a fresh impetus for continuing the search of magnetostrictive materials suitable for developing transducers. Clark [19] has obtained room-temperature magnetostriction in the alloy of Tb and Fe, which also has higher Curie temperature (around 700 K). Subsequently, it is found that by adding another rare earth material called dysprosium with Tb–Fe alloy, the magnetic anisotropy in the alloy can be reduced, thus generating even larger strains. The commercially available and well-known magnetostrictive material Terfenol-D is an example of the aforementioned alloy of terbium, iron, and dysprosium. However, the proportion of Tb, Fe, and Dy varies depending on specific requirement of magnetoelasticity and temperature characteristics. For example, in reference [20], it is shown experimentally that the 'Terfenol' compound made using the composition $Dy_{0.73}Tb_{0.27}Fe_{1.95}$ produces less free strain than the same compound made of $Dy_{0.7}Tb_{0.3}Fe_{1.95}$ composition. However, in the former, the strain varies more linearly with the magnetic field as compared to the latter. This makes the first alloy more suitable for actuation purpose. Also, substituting dysprosium from Tb–Fe alloy by other rare earth materials like holmium or samarium, magnetostriction characteristics can be significantly changed.

In the linear region of actuation and sensing (assuming the vector quantities along the direction of maximum response), the constitutive equations of magnetostriction are given by

$$S = s^H \sigma + dH, \tag{7.6.1}$$

$$B = d\sigma + \mu^T H, \tag{7.6.2}$$

where S is the strain, σ the mechanical stress, H the magnetic field intensity, and B the flux density. The compliance value at a constant magnetic intensity is denoted by s^H, d is the magnetomechanical constant, and μ^T is the permeability of the medium under constant stress condition. The equations are of the same form as that of piezoelectricity. However, the study of any typical S–H and B–H curve [20] brings out the following observations.

The relationship between magnetostriction and applied magnetic field is highly dependent on the intensity of the magnetic field. The relationship is approximately linear when the intensity of the applied magnetic field H is much lower than the intensity of the polarizing field H_{pol} (field at which the magnetic domains are initially aligned). The nonlinearity begins as H approaches H_{pol} and the curve gradually flattens out signifying saturation or completion of all the domain alignments. Typically, for Terfenol-D rods under stress-free condition, such a relationship is approximately linear in the range of magnetic field from 0 to 100 Orsted.

The maximum free strain generated by magnetostriction is quite large, almost twice as much as that of PZT. Yet, unlike piezoelectric material, the reversal of magnetic field does not result in the reversal of strain here. Particularly, for dynamic applications like vibration suppressions, reversal of actuation strain is very much necessary. Hence, for such applications, these actuators are operated with a biased magnetic field such that with respect to the biasing center, reversal of strain occurs. The technique, however, reduces the availability of actuation strain by approximately 50%, thus lowering its edge over the piezoelectric materials.

It is observed that hysteresis is present in the B–H curve and is usually absent in the B–S curve. As a result, the combination of the two curves shows hysteresis in the S–H curve. The extent of hysteresis depends on the stoichiometry of the material and prestressing on the actuator.

A comparison of the material properties of various available magnetostrictive materials is shown in Table 7.4. Apart from Terfenol-D, the other smart materials included in the list are some of the more recent varieties in this field. Metglas or Vitrovac is a commercially available compound of iron (Fe), silicon (Si), and boron (B). The other two materials consist of composites of Terfenol powders mixed with nonmetallic binders [21]. The binders work as insulating layers and effectively reduce the eddy current loss at higher frequencies of the applied magnetic field. The isotropic and anisotropic properties are attributed to random or biased magnetic

Table 7.4 Salient properties of magnetostrictive materials and aluminum (for reference)

Material properties	Terfenol-D	Metglas	Isotropic composite	Anisotropic composite	Aluminum
E_m (GPa)	25–35	55–60	19.7	17.0	73.0
ρ (kg/m^3)	9250	7470	7500	6800	2630
d (nM/A)	15.0	–	3.7	5.9	0
Λ (μ-strain)	1500	52	400	630	0

dipole orientation of Terfenol powders during manufacturing. Table 7.4 shows that Terfenol-D rod produces the highest free strain Λ. It is considered to be the most effective material from the point of view of actuation strain generation. However, it has poor tensile strength, and it is the heaviest among the magnetostrictive materials. On the other hand, Metglas possesses better mechanical properties, although it generates the lowest actuation strain. It is used mostly as a sensing material. In the two composite magnetostrictive materials, the energy conversion efficiency is quite low ($k = 0.28$ and 0.40, respectively) rendering these unsuitable for low-frequency applications. Magnetostrictive materials with the capacity of high strain generation and wide bandwidth of operation promise wider applications in smart structures in near future.

7.7 Magnetostrictive Sensors and Actuators

We have noted in Table 7.2 that for smart structural applications magnetostrictive materials, like Terfenol-D, are as good as piezoelectric materials. The free strain or the maximum strain under no-load condition in Terfenol-D is about 2000 μ-strain, which is comparable to free strain of polycrystalline piezoelectric material; the bandwidth of the material is up to MHz level indicating its superiority over SMA. Also, because of higher modulus of elasticity, the energy density of Terfenol-D is maximum among all the smart materials. In this section, we will discuss how one can utilize these properties in developing actuators and sensors.

Many researchers have developed mini- and micro-actuators using Terfenol-D. Consider, for example, the mini-actuator developed by Anjannappa and Bi [22]. A typical magnetostrictive mini-actuator is shown in Fig. 7.6.

It may be noted that the Terfenol-D rods work very much similar to piezo-stacks. The governing equation describing the constrained strain is similar to Eq. (7.4.5). However, unlike piezo-actuators, the effect of thickness of a magnetostrictive mini-actuator (MMA) is not negligible. Hence, the induced (constrained) strain is related to the free strain by the following relationship:

$$\varepsilon_c = \left(1 + \frac{Ebt^2}{12A_m E_m (t + t_m)}\right)^{-1} \Lambda, \tag{7.7.1}$$

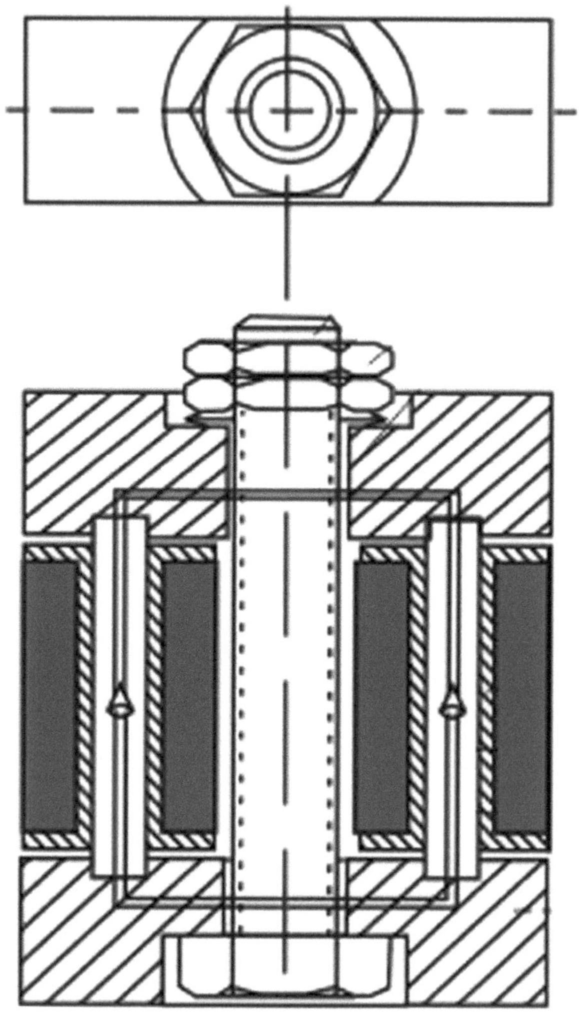

Fig. 7.6 A magnetostrictive mini-actuator

where b and t are, respectively, the width and thickness of the host beam, A_m and t_m are the cross-sectional area and thickness of the MMA. E denotes the modulus of elasticity of the host beam, and E_m denotes the modulus of elasticity of the magnetostrictive material. The expression for the free strain Λ contains an additional term—the thermal effect due to current passing through the solenoid, and hence, Λ is expressed as

$$\Lambda = \mathrm{d}Gi(t) + \alpha' K \int_0^t \mathrm{e}^{-t/C_2} i^2(t)\,\mathrm{d}t, \tag{7.7.2}$$

where G is the coil constant, α' is the coefficient of equivalent thermal expansion, K and C_2 are two thermal constants which are obtained experimentally, and i is the current passed through the magnetizing coil.

Example 7.7.1 Consider the same MEMS beam similar to that discussed in the Example 7.4.1. However, instead of piezoelectric layers on top and bottom, the host beam is excited by a magnetostrictive mini-actuator of same length as that of the host beam. The MMA has a cross-sectional area of 4×10^{-8} m^2 and thickness of 100 μm. The elastic modulus of the magnetostrictive actuator is 50 GPa, the magnetomechanical constant is 15 nM/A, and a current of 750 mA is passed through the coil. Considering the coil constant G to be 10,000 and neglecting the thermal effect, find out the strain induced at the host beam.

Solution: Neglecting the thermal effect, the free strain that could be developed by the MMA could be obtained using Eq. (7.7.1) as $\Lambda = 15 \times 10^{-9} \times 10000 \times 750 \times 10^{-3} = 112.5\,\mu$-strain.

Using Eq. (7.7.1), the strain induced on the host beam is

$$\varepsilon_c = \left(1 + \frac{100 \times 10^9 \times 200 \times 10^{-6} \times 100 \times 10^{-6}}{12 \times 4 \times 10^{-8} \times 50 \times 10^9 \times 200 \times 10^{-6}}\right)^{-1} \times 112.5 \times 10^{-6} = 0.27\,\mu\text{-strain}.$$

Magnetostrictive sensors are developed based on three phenomena: (i) the Villari effect, which refers to the changes in magnetization when a magnetostrictive material is subjected to applied uniaxial stress; (ii) the Matteucci effect, referring to the changes in axial magnetization of a current carrying amorphous wire when it is twisted; (iii) the change in permeability of these materials when subjected to applied stress. Magnetostrictive delay line sensors are based on this phenomenon. Here, we will discuss the principles of sensors developed based on the Villari effect.

The principle linked with the magnetization and voltage pickup in a coil is the well-known Faraday–Lenz law of electromagnetic induction given by

$$V = -NA\,\frac{\mathrm{d}B}{\mathrm{d}t} \tag{7.7.3}$$

where N is the number of turns of the pickup coil, and A is the cross-sectional area of the coil. The magnetic induction B is obtained using Eqs. (7.6.1) and (7.6.2) as

$$B = d\sigma + \mu\left(S - s^H\sigma\right)/d = \frac{\mu}{d}S + \left(d - \mu\frac{s^H}{d}\right)\sigma \tag{7.7.4}$$

Substituting Eq. (7.7.4) into (7.7.3), we can get the sensed voltage

$$V = -NA\left[\frac{\mu}{d}\frac{\mathrm{d}S}{\mathrm{d}t} + \left(d - \mu\frac{s^H}{d}\right)\frac{\mathrm{d}\sigma}{\mathrm{d}t}\right]$$

For harmonic excitation of frequency ω, Eq. (7.7.4) may be written as

$$V = -NA\,(2\pi\omega)\left[\frac{\mu}{d}S_0 + \left(d - \mu\frac{s^H}{d}\right)\sigma_0\right], \tag{7.7.5}$$

where S_0 and σ_0 are amplitudes of strain and stress. It has been observed from experiments that the first term of Eq. (7.7.5) is quite small in comparison with the last two terms. Hence, Eq. (7.7.5) could be simplified by neglecting this term and used for force sensing ($F = \sigma_0 A$) as

$$F = \frac{V}{2\pi\omega N}\left[\frac{1}{d - \mu\frac{s^H}{d}}\right] \tag{7.7.6}$$

Magnetostrictive sensor could also be used for strain and displacement sensing, and the details of such systems will be discussed in subsequent chapters.

7.8 Active Fiber Composites

From Table 7.3, it may be noted that the magnitude of piezoelectric constant d_{33} is significantly higher than d_{31}. For example, for a soft piezoelectric material, d_{31} is about −55 pC/N, whereas d_{33} is around 190 pC/N. Similarly, for a hard piezoelectric material, d_{33} is 2.5 times higher than d_{31}. Internally leveraged piezoelectric actuators like piezo-stacks are developed exploiting this property. This has also tempted material scientists to develop piezoelectric fibers and actuate them axially to generate larger actuation strain. Later, these fibers are embedded inside a polymeric matrix to form composite actuator and sensor. Two types of smart composites have been developed using piezo-ceramic fibers; these are active fiber composite (AFC) and macro-fiber composite (MFC). The difference between AFC and MFC is in the manufacturing process of the fiber. While the AFC fibers are developed using standard sol–gel technique, the MFC fibers are essentially chopped from PZT blocks. The MFC fibers are rectangular in cross section, and hence, it offers better electrical contact between the fibers.

The model of a typical AFC lamina having interdigitated electrode (IDE) surface is shown in Fig. 7.7. Following the balance of active and reactive moments similar to Eq. (7.4.4), the active strain at a distance z from the neutral axis can be expressed as [23]:

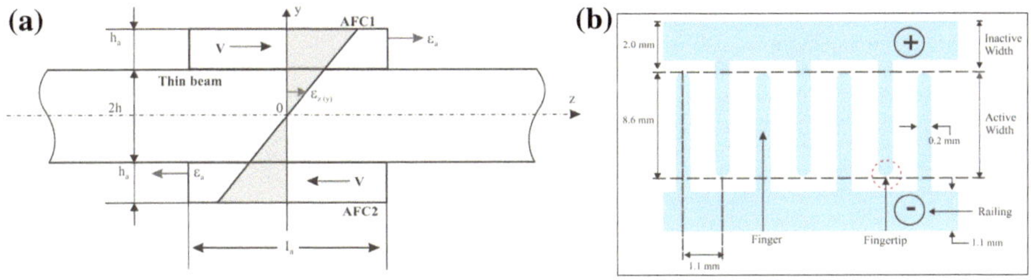

Fig. 7.7 Active fiber composite: **a** stress distribution across the thickness and **b** interdigitated poling of the fibers

$$\varepsilon_x = K\Lambda z,$$
$$K = \frac{3E_a\left[(h+h_a)^2 - h^2\right]}{2\left[E_a\left\{(h+h_a)^3 - h^3)\right\} + Eh^3\right]} \quad \text{and} \qquad (7.8.1)$$
$$\Lambda = d_{33}\frac{V}{h_a}.$$

One may note from Fig. 7.7b that the voltage is applied along the length of the fiber, and hence, the free strain Λ is obtained using the piezoelectric constant d_{33}. This is the major difference between the actuation through a conventional piezoelectric patch as shown in Fig. 7.7a, where the voltage is applied out of the plane. In order to efficiently develop this voltage, the interdigitated arrangements of electrode shown in Fig. 7.7c–d are used.

7.9 Shape Memory Alloy (SMA)-Based Actuators

Shape memory effect (SME) is the property of shape memory alloy (SMA) by virtue of which these can recover apparent permanent strains to revert back to its original shape when they are heated above a certain temperature. SMAs have two stable phases—the high-temperature phase called austenite (parent phase) and the low-temperature phase called martensite. Martensite has monoclinic crystal structure, while austenite has cubic crystal structure. SME occurs because of phase transformation of martensite to austenite beyond recrystallization temperature.

In the absence of any applied load, upon heating an SMA, the martensite phase simply gets transformed to austenite. During this phase transformation, though there is change in the microstructure of SMA, macroscopic shape of the SMA does not change. Again, on cooling the SMA, a reverse martensite transformation takes place; however, the austenite now gets transformed to twinned martensite. This transformation is called self-accommodating transformation [24]. Figure 7.8 shows the phase transformations in SMA.

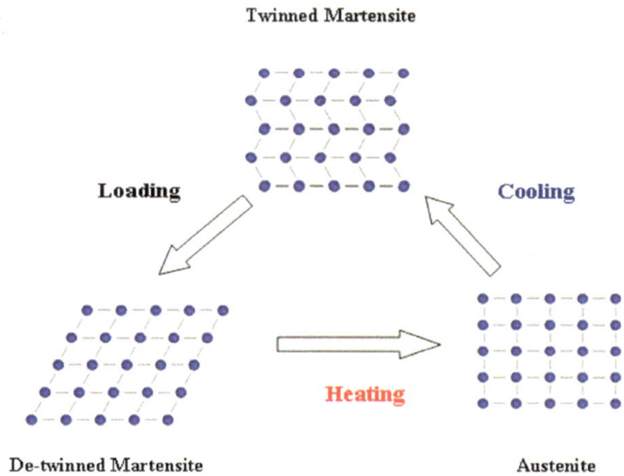

Fig. 7.8 Phase transformation in SMA

Now, upon loading SMA in the martensite phase, de-twinning of martensite takes place resulting in large straining of SMA. The strain produced in SMA during complete transformation of twinned martensite to de-twinned martensite is known as transformation strain [Λ_t] and is the maximum strain that SMA can recover upon heating. Further loading of SMA causes permanent strain which cannot be recovered. Recovered transformation strain of SMA is directly proportional to the degree of transformation of twinned martensite to de-twinned martensite.

There are four critical temperatures for SMA which characterize its behavior. These are A_s, A_f, M_s, and M_f with A and M indicating austenite and martensite phases. The subscripts, s and f, denote start and finish of transformation, respectively. These temperatures are stress dependent, and their values change depending on the loading conditions.

The shape memory effect (SME) can be summarized as shown in Fig. 7.9. At a temperature below M_{f0}, SMA has twinned martensitic microstructure (B). On loading SMA, the de-twinning of martensite starts and it ends at some critical value of the stress (C). Further loading of SMA results in permanent plastic strain in SMA. On complete unloading, SMA still retains strain in it (D), which is recovered if it is heated beyond A_{f0} (E). At A_{s0}, martensite phase starts transforming to austenite. A_{f0} is the temperature for completing this transformation at zero stress level. Percentage of transformation strain recovered by SMA is proportional to austenite volume fraction. On cooling SMA, martensite transformation starts from M_{s0}, and for temperatures below M_{f0}, austenite gets completely transformed to twinned martensite (B). C_m and C_A are material-dependent parameters of SMA providing the relationship between temperature and stress for martensite and austenite phases, respectively. Their units are Pa/K. C_M is typically of higher magnitude than C_A.

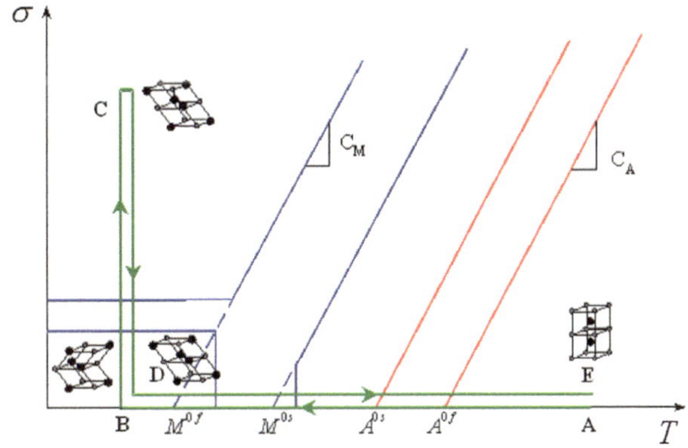

Fig. 7.9 SMA stress–strain behavior

 — Austenite phase with cubic crystal

 — Twinned martensite phase with orthorhombic crystal

 — De-twinned martensite phase.

SME constitutes both stress-induced transformation (twinned martensite to de-twinned martensite) and temperature-induced transformation (de-twinned martensite to austenite). SME is a useful property of SMA for applications in the field of linear actuators as the displacement range of an actuator can be easily controlled by controlling the temperature of the SMA element. However, to model the SME, we can use micro-mechanics approach, which considers martensitic transformation as inclusion, and obtain stress–strain relationship based on free-energy estimation. Alternately, we can develop a phenomenological model based on experimental data. These may be determined based on stress–temperature diagram. Tanaka first proposed a convenient phenomenological model [25]. This model holds good for one-dimensional SMA and is obtained from general three-dimensional theory based on energy balance equation. The model is based on three state variables, which are strain (ε), temperature (*T*), and martensite volume fraction (ξ), respectively. The relationship is given by following first-order differential equation:

$$\dot{\sigma} = E\dot{\varepsilon} + \theta\dot{T} + \Omega\dot{\xi}, \tag{7.9.1}$$

where σ is the Piola–Kirchhoff stress, ε is the Green strain, E is the Young's modulus, θ is the thermal expansion coefficient, and Ω is the transformation modulus. The dot at the top indicates differentiation with respect to time. Further, the de-twinned martensite volume fraction (ξ) could be expressed as exponential function of stress and temperature as given below.

For austenite to martensite transformation

$$\xi = 1 - \exp\left[a^M(M_s - T) + b^M\sigma\right] \tag{7.9.2}$$

where, $a^M = \frac{\ln(0.01)}{M_s - M_f}$ and $b^M = \frac{a^M}{C_M}$.

For martensite to austenite transformation

$$\xi = 1 - \exp\left[a^A(A_s - T) + b^A\sigma\right], \tag{7.9.3}$$

where $a^A = \frac{\ln(0.01)}{A_s - A_f}$ and $b^A = \frac{a^A}{C_A}$.

The slopes C_M and C_A are indicated in Fig. 7.9

Tanaka's model describes the behavior of SMA qualitatively and faces problems like singularity in the martensitic fraction model. Liang and Rogers [26] improved this model by replacing the exponential with the cosine function. Brinson [27] further improved Liang's model by introducing the fractions of stress-induced and temperature-induced martensite volume fraction as

$$\xi = \xi_S + \xi_T, \tag{7.9.4}$$

where ξ_S denotes the fraction of stress-induced single-variant de-twinned martensite, and ξ_T is the fraction of temperature-induced multiple variant twinned martensite phases. The cosine model is commonly referred as Liang–Brinson model and is given below.

Conversion to de-twinned Martensite:

For $T > M_S$ and $\sigma_S^{cr} + C_M(T - M_S) < \sigma < \sigma_f^{cr} + C_M(T - M_S)$

$$\begin{aligned} \xi_S &= \frac{1 - \xi_{S0}}{2}\cos\left\{\frac{\pi}{\sigma_S^{cr} - \sigma_f^{cr}}\left[\sigma - \sigma_f^{cr} - C_M(T - M_S)\right]\right\} + \frac{1 + \xi_{S0}}{2} \text{ and} \\ \xi_T &= \xi_{T0} - \frac{\xi_{T0}}{1 - \xi_{S0}}(\xi_S - \xi_{S0}) \end{aligned} \tag{7.9.5}$$

The subscript '0' indicates the initial fraction of transformation. The superscript 'cr' indicates the critical stress beyond which the multivariant martensite crystals get converted to single variant of martensite.

For $T < M_S$ and $\sigma_S^{cr} < \sigma < \sigma_f^{cr}$

$$\xi_S = \frac{1 - \xi_{S0}}{2}\cos\left\{\frac{\pi}{\sigma_S^{cr} - \sigma_f^{cr}}\left(\sigma - \sigma_f^{cr}\right)\right\} + \frac{1 + \xi_{S0}}{2}$$

and

$$\xi_T = \xi_{T0} - \frac{\xi_{T0}}{1 - \xi_{S0}}(\xi_S - \xi_{S0}) + \Delta_{T\xi}$$

where if $M_f < T < M_s$ and $T < T_0$

$$\Delta_{T\xi} = \frac{1 - \xi_{T0}}{2}\{\cos[a_M(T - M_S)] + 1\} \text{ else } \Delta_{T\xi} = 0$$

Conversion to austenite:
For $T > A_S$ and $C_A(T - A_f) < \sigma < C_A(T - A_S)$

$$\xi = \frac{\xi_0}{2}\left\{\cos\left[a_A(T - A_S) - \frac{\sigma}{C_A}\right] + 1\right\} \tag{7.9.6}$$

$$\xi_s = \xi_{s0} - \frac{\xi_{s0}}{\xi_0}(\xi_0 - \xi) \text{ and } \xi_T = \xi_{T0} - \frac{\xi_{T0}}{\xi_0}(\xi_0 - \xi)$$

With separation of ξ into two distinct components, Liang's constitutive model is further improved, to consider material functions, as

$$\sigma - \sigma_0 = E(\xi)\varepsilon - E(\xi_0)\varepsilon_0 + \Omega(\xi)\xi_s - \Omega(\xi_0)\xi_{s_0} + \Theta(T - T_0), \tag{7.9.7}$$

where $E(\xi) = E_A + \xi(E_M - E_A)$; $\Omega(\xi) = -\varepsilon_L E(\xi)$; E_A is the Young's modulus of the austenite; E_M is the Young's modulus of the martensite; and ε_L is the maximum recoverable martensitic strain.

7.10 Electrorheological and Magnetorheological Fluid-Based Dampers

Electrorheological fluids (ERF) and magnetorheological fluids (MRF) are characterized by a colloidal suspension of electrically/magnetically semiconductive particle in an insulating carrier fluid. Silicone oil is commercially used as the carrier fluid having carbonaceous particles in suspension. The particle size is typically of the order of microns. The ERF exhibit a dramatic change in viscosity upon application of external electric field due to the polarization of the suspended particles between the electrodes. The gap between the electrodes is of the order of millimeters. In MRF, a drastic change in viscosity occurs due to the application of a magnetic field. The rapid and reversible phenomenon of change in viscosity against electric field was first observed by Winslow in 1949 and often referred to as 'Winslow effect.' The flow behavior of ERF is generally modeled as Bingham fluid, in which the shear stress τ, upon application of an electric field E, is given by

$$\tau = \tau_y + \eta \, \dot{\gamma} \text{ and } \tau_y \propto E^{\alpha} \qquad (7.10.1)$$

where $\dot{\gamma}$ is the shear strain rate, η is the plastic viscosity, and τ_y is the yield stress induced by the electric field with α close to 2 for a moderate value of E. ERF are commonly used for vibration isolation. A typical use of ERF for vibration isolation is shown in Fig. 7.10 [28]. The vibration isolator consists of a parallel combination of a helical spring and a piston–cylinder dashpot with ERF. It has been observed that with suitable combination of particle concentration, carrier fluid and the strength of the electric field the damping characteristics of the dashpot can be varied over a wide range. At high field strength, even Coulomb damping characteristic (rather than viscous) can be achieved.

ERF can be used for both passive vibration isolation and active vibration isolation. In case of passive isolation, a fixed field could be applied to achieve Coulomb damping characteristics. However, for optimal use, over a broad frequency range, one can actively control the electric field to the ERF mount for vibration control. In Fig. 7.11, a typical application of active ERF mounting on a flexible beam is shown [29]. Here, the flexible beam is excited by an

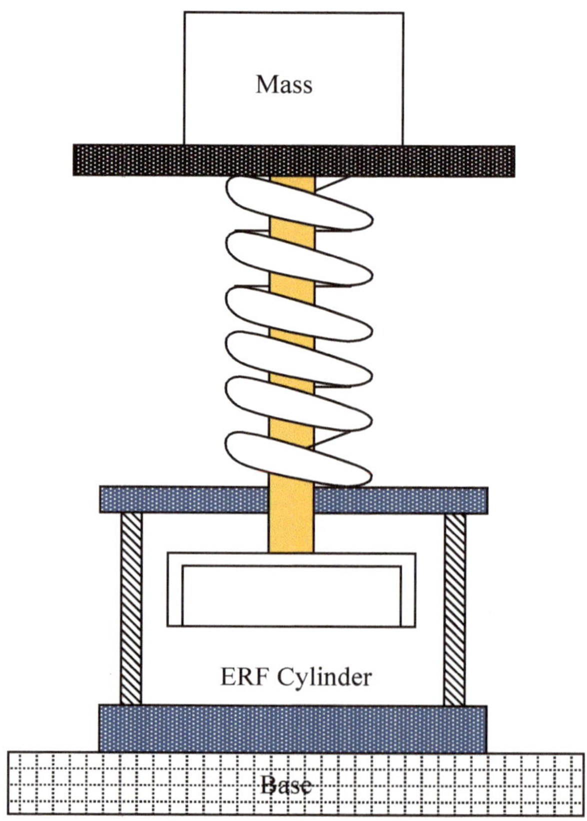

Fig. 7.10 A typical vibration isolator with electrorheological fluid

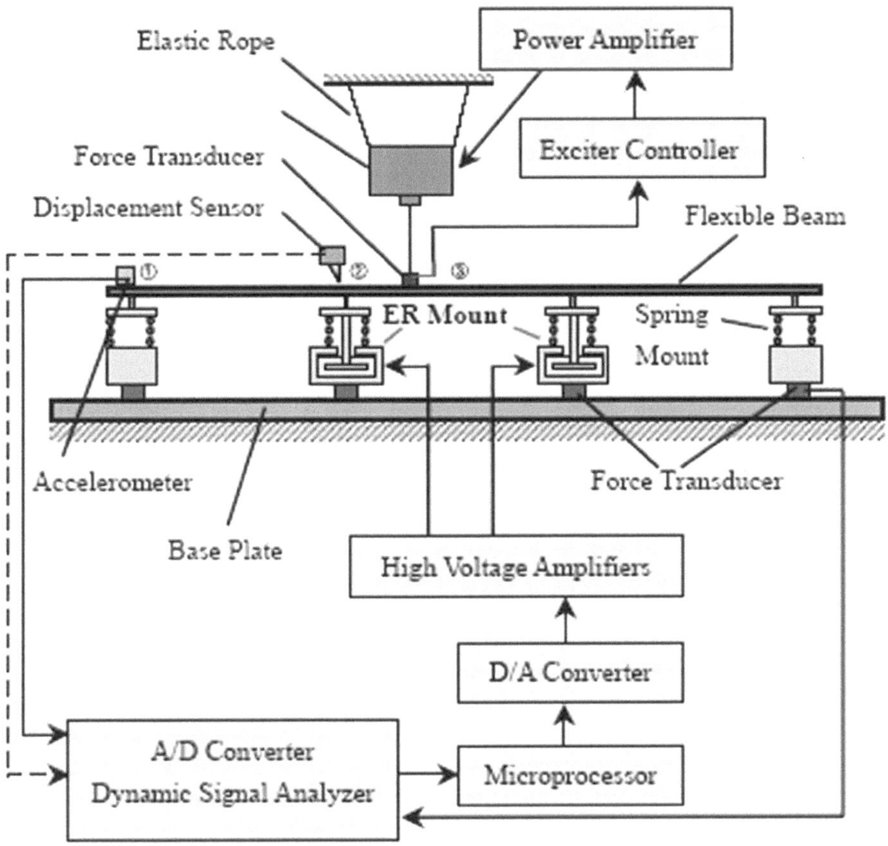

Fig. 7.11 Application of ER fluid-based actuator

electromagnetic exciter; acceleration, displacement, and force are sensed at locations 1, 2, and 3, respectively, and fed back to a microprocessor, which in turn controls the voltage applied to the ERF mount.

7.11 Some Recent Developments

New actuators are being developed based on phenomena like flexing of polymeric chains during glass transition (shape memory polymer (SMP)) and charge migrations in polymers like electroactive polymers (EAP). Based on some phenomenological models of interaction between chemical, electrical, and mechanical fields, a relationship between induced free strain and applied voltage in EAP is obtained. This expression of free strain could then be used in Eq. (7.4.11) or (7.8.1) to obtain the induced strain, for embedded or surface-bonded actuation [30]. For SMP, an approach similar to SMAs is used, though the constitutive relationship differs significantly from the SMA [31].

References

1. Breitbach EJ, Lammering R, Meleher J, Nitzsche F (1994) Smart structures research in aerospace engineering. In: Proceedings of SPIE second European conference on smart structure and materials, vol 2361
2. Newnhan RE, Ruschan GR (1991) Smart electronic systems. J Am Ceram Soc 74:463–480
3. Coghlan A (1992) Smart ways to treat materials. New Sci 13:27–29
4. Bunk WGJ (1991) Advanced structural and functional materials. Springer, Berlin
5. Stanway R, Sporston JL (1994) Electro-rheological fluids, a systematic approach to classify modes of operation. Trans ASME J Dyn Syst Meas Control 504:116–198
6. Skinner DP, Newnham RE, Cross LE (1978) Flexible composite transducers. Mater Res Bull 13:599–607
7. Crawley EF (1994) Intelligent structures for aerospace: a technology overview and assessment. AIAA J 32(8):1689–1699
8. Baz A, Poh S, Ro J, Gilheany J (1995) Control of the natural frequencies of NiTiNOL reinforced composite beams. J Sound Vib 185(1):171–185
9. Srinivasan AV, Gutts DG, Schetky LM (1991) Thermal and mechanical considerations in using shape memory alloys to control vibration in flexible structures. Metall Trans A 22A:623–627
10. Measures RM (1989) Smart structures with nerves of glass. Aerosp Sci 26:289–351
11. Kordonsky WI (1993) Magnetorheological effect as a base of new devices and technology. J Magn Magn Mater 122:395–398
12. Berlincourt D (1981) Piezoelectric ceramics: characteristics and applications. J Acoust Soc Am 70(6):1586–1595
13. Crawley EF, Luis JD (1987) Use of piezoelectric actuators as elements of intelligent structures. AIAA J 25(10):1371–1385
14. Sessler GM (1981) Piezoelectricity in Polyvinylideneflouride. J Acoust Soc Am 70(6):1567–1576
15. Niezrecki C, Brei D, Balakrishnan S, Moskalik A (2001) Piezoelectric actuation: state of the art. Shock Vib Dig, 269–280
16. Crawley EF, Anderson EL (1991) Detailed models of piezoceramic actuation of beams. J Intell Mater Syst Struct 1(1):4–25
17. Dosch JJ, Inman DJ, Garcia E (1992) A self-sensing piezoelectric actuator for collocated control. J Intell Mater Syst Struct 3:166–183
18. Vipperman JS, Clark RL (1994) Implementation of an adaptive piezoelectric sensoriactures. J Acoust Soc Am 96(1):294–299
19. Clark AE (1980) Magnetostrictive rare earth Fe compounds. In: Ferromagnetic materials, Chap. 7
20. Greenough RD, Jenner AG, Schulze MP, Wilkinson AJ (1991) The properties and applications of magnetostrictive rare-earth compounds. J Magn Magn Mater 101:75–80
21. Sandlund L, Fahlander M, Cedell T, Brestorff JB, Clark AE (1994) Magnetostriction elastic moduli and coupling factors of composite Terfenol-D. J Appl Phys 75:5656–5658
22. Anjanappa M, Bi J (1994) Magnetostrictive mini actuators for smart structure applications. Smart Mater Struct 3:383–390
23. Nguyen C, Kornmann X (2006) A comparison of dynamic piezoactuation of fiber-based actuators and conventional piezo patches. J Intell Mater Syst Struct 17:45–56

24. Otsuka K, Wayman CM (1998) Shape memory materials. Cambridge University Press, Cambridge
25. Tanaka K (1986) A thermomechanical sketch of shape memory effect: one dimensional tensile behavior. Res Mechanica 18:251–263
26. Liang C, Rogers CA (1997) One dimensional thermomechanical constitutive relations for shape memory materials. J Intell Mater Syst Struct 8:285–302
27. Brinson LC (1993) One dimensional constitutive behavior of shape memory alloys: thermomechanical derivation with non-constant material functions. J Intell Mater Syst Struct 4:229–242
28. Tandon I, Mallik AK, Gupta Bhaya P (1999) Performance characteristics of a vibration isolator with electro-rheological fluids. J Sound Vib 219(3):395–404
29. Jung WJ, Jeong WB, Hong SR, Choi SB (2004) Vibration control of a flexible beam structure using squeeze-mode ER mount. J Sound Vib 273:185–199
30. Bandopadhya D, Bhattacharya B, Dutta A (2007) Active vibration control strategy for a single link flexible manipulator using ionic polymer metal composite. J Intell Mater Syst Struct (On line version July 10, 2007)
31. Tobushi H, Okumura K, Hayashi S, Ito N (2001) Thermo-mechanical constitutive model of shape memory polymer. Mech Mater 33:545–554

Chapter 8
Smart Sensors for Structural Health Monitoring

8.1 Introduction

In the introductory chapters, we have discussed various causes that can create damages in pipelines. These include chemical and stress corrosion, sudden variation in gravity and thermal loading, flow-induced vibration, impact damage at the pipe joints. We have also discussed how different sensors can be used for sensing these damages. Optical and Laser Scanning, Magnetic Flux Leakage Analysis, Microwave and Ultrasonic Sensing are some of the techniques discussed in the overview. However, most of the above-mentioned sensors generate high density of data creating challenges for data storage as well communication. Smart material-based sensors like PVDF sensing are more advantageous in such cases. PVDF sensor is developed from a subset of polymeric piezoelectric material; the nature and constitutive relationship of such smart materials are discussed at length in the earlier chapter. In this chapter, we will more focus on the use of such smart materials in developing structural health monitoring (SHM) System. As a background to this discussion, it will be important for us to briefly review the concept and scope of SHM.

SHM has been defined and discussed extensively in the last two decades due to enormous advancements in smart material-based sensing technology. In a broad sense, this refers to the implementation of a sensing strategy based on distributed smart sensors with the following four sequential objectives:

(a) Detect the presence of damage, if affirmative then;
(b) Find out the location of damage;
(c) Estimate the extent of damage and finally;
(d) Interpret the nature of damage and estimate the remaining useful life.

While the first two functions are generally sub-grouped as **damage identification**, the last two are commonly known as **damage prognosis**.

The following are the basic axioms of SHM proposed by the researchers [1].

H. Ogai and B. Bhattacharya, *Pipe Inspection Robots for Structural Health and Condition Monitoring*, Intelligent Systems, Control and Automation: Science and Engineering 89, https://doi.org/10.1007/978-81-322-3751-8_8

Axiom I: Every material has inherent flaws or defects

Generally, engineering materials are classified into four major groups—metals, polymers, ceramics, and composites. Among these, except metals, all other materials are more susceptible to damages due to inherent weakness in the materials at the microscopic level. For metals, the defects may accrue due to extreme environments, fatigue, corrosion, etc.

Axiom II: The process of damage identification requires a comparison between two system states

Without the availability of normal response profile from a system, and subsequently comparing the same with changed system response, it is impossible to ascertain the presence and extent of damage. In addition, for damage prognosis, one needs response data related to various damage cases as well to develop the knowledge base.

Axiom III: While damage existence and location can be obtained in an unsupervised learning mode, damage prognosis can generally only be done in a supervised learning mode

This axiom is an extension of Axiom II, where, we have mentioned that for *damage prognosis*, knowledge base development corresponding to various kinds of damages is essential. This is further used for supervised learning.

Axiom IVa: Sensors alone cannot measure damage. Feature extraction through signal processing and classification is necessary to transform sensor data into meaningful damage information

Also:

Axiom IVb: Without intelligent feature extraction, the more sensitive measurement is to damage, the more sensitive it is to changing operational and environmental conditions

This axiom emphasizes that a mere increase in the sensitivity of a sensor is not sufficient for damage detection, unless suitable features related to damaged system based on engineering judgment are chosen.

Axiom V: The length scale and timescale associated with damage initiation and growth of damage dictate the required properties of the SHM sensing system

There are certain damages like low cycle fatigue cracks which occur at a slower rate in comparison to damages like fast fracture in ceramics. The effect of length scale is material dependent—ceramics and polymers show defects in macroscale while metals generally contain defects at micro/nanoscale level. Hence, the choice of sensors should be made with a proper judgment of the expected type of damage in the system and its' material property set. This axiom guides us in terms of selection of sensors. Keeping the two scales in mind, one needs to decide on: nature and type of data acquisition, number and location of sensors, sensor bandwidth, etc.

Axiom VI: There is a trade-off between the sensitivity to damage of an algorithm and its noise rejection capability

It is generally observed that if the signal-to-noise ratio increases, the threshold level of damage detection worsens. To improve the scenario, one needs to employ techniques like digital wavelet transformation along with suitable filtering of the data.

Axiom VII: The size of damage that can be detected from changes in system dynamics is inversely proportional to the frequency range of excitation.

This is true especially for ultrasonic devices, where, higher frequency implies smaller wavelength. Usually it is considered that if the flaw size in a specimen is greater than half of the wavelength of the ultrasonic signal, the flaw will be detectable. Thus, smaller wavelength will be required in detecting smaller flaws in a structural system.

The SHM is envisaged to be useful in increasing the safety of operation, improving reliability, and also expanding the life of a product. In comparison to the traditional sensing systems, SHM relies on low-cost smart sensors, reduces the dependence of human-in-loop support and specifically designed for Condition-based Monitoring as opposed to Schedule-based Maintenance. Table 8.1 provides a brief summary of advantages and disadvantages of Condition-based Monitoring.

Various stratifications of functions to be performed to implement SHM are shown in Fig. 8.1.

As depicted in Fig. 8.1, SHM involves the observation of the dynamic response of a system over time using periodically sampled measurements from an array of sensors. The measurements can be related to mechanical deformation (strain/displacement) or its rate of change, stress, temperature, moisture level, etc. It also involves developing algorithms for the extraction of damage-sensitive features

Table 8.1 Advantages and disadvantages of condition-based monitoring

Advantages of condition-based monitoring	Disadvantages of condition-based monitoring
As this is performed while a system is in operational state, it causes minimum disruption	It requires large investment in the beginning for developing sensor networks
It improves work safety and reduces the cost of sudden failure	It also needs more trained/knowledgeable people to run a system
It increases reliability of the equipment	It may require system modifications to retrofit the system with sensors
It reduces the chances of unscheduled down-time and maintenance and thus significantly reduces the operating costs	It is relatively less sensitive to uniform degradations like wear and fatigue
It also reduces the chances of collateral damages in a system which may be critical for systems containing many complex sub-systems	

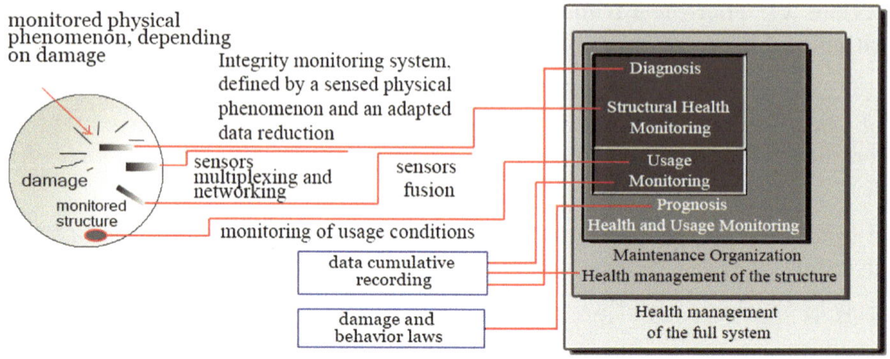

Fig. 8.1 A typical structural health monitoring system [2]

from these measurements, and subsequently the statistical analysis of these features to determine the current state of health of a system. In terms of industrial application, SHM technology has been successfully applied in four broad domains: Aerospace Systems, Civil Infrastructures like Bridges, Transportation systems like Railroads and Underground transportation, Power generation systems like Wind turbines and Nuclear Power Plants. Application of SHM in Automobiles is also rapidly coming up, while the utility of the same for consumer products is considered to be an active area of research today. In the following section, we will provide a brief description of these applications.

8.2 Industrial Applications of SHM

As mentioned earlier, one of the foremost applications of SHM is in the field of Aerospace Systems [3]. This has happened since aircraft industry spends close to thirty percent of life cycle cost on inspection of aircraft. It is noted that by successful implementation of SHM approximately 40% of time of inspection has been saved for fighter aircraft. One of the biggest beneficiaries of SHM in Aerospace applications is in the Helicopter sector. The establishment of Health and Usage Monitoring System (HUMS) for main rotor and gear box system of large rotorcrafts has significantly increased the safety of the Helicopters. In the last ten years, there is also a change in the business model in many industries where the aero-engines are leased by the manufacturer instead of complete sale. In this model, the manufacturer keeps track of the condition of the engines and thus guarantees both safety as well as optimal utilization of the system.

In the early 80s, SHM was introduced in the offshore oil rigs where the systems are subjected to highly unpredictable aero and hydrodynamic loading. A strategy of vibration-based condition monitoring was proposed which intended to identify and localize the presence of damage by sensing the change in natural frequency of the system. Although, the proposed system was not widely accepted in the oil industry for issues like noise control, difficulty of sensor embedding and degradation of electronic systems due to corrosive environment; the same has been subsequently adopted by the rotor industry and civil infrastructure maintenance industry extensively. Hence, today, the key application of SHM is in the field of vibration sensing and control.

8.3 SHM for Vibration Sensing and Control

Distributed sensing and control of vibrating structural systems like beams and plates have been studied since the beginning of the 80s. The major application of this concept was in terms of trajectory control of slender aerospace structures like rockets and missiles by controlling the thrusts from the jets. The stability margins for such systems while subjected to distributed follower force tangential to the deformed body are enhanced by successful applications of SHM. With the advent of smart materials, this technique has gained more practical utility as truly distributed sensing and actuation layers can be developed from smart materials. Piezoelectric polymers like Polyvinylidene Fluoride (PVDF) are quite compliant and could be applied in patches over any curved surface. PVDF film-based distributed actuators and sensors have been designed for vibration attenuation in beams and plates [3]. The technique is known as spatial shading in which biaxial PVDF films are placed in ramp configuration to develop a single all-mode vibration controller. A typical configuration is shown in Fig. 8.2. The base plate may be either isotropic material like metals or anisotropic composite material like Carbon-fiber-reinforced composite.

In both the cases, the addition of smart layer over or under a base plate makes it a composite structure. Also, laminated composite materials are being used today extensively for piping. In the following section, we will introduce the SHM of composite beam structure as a generic case. The ideas developed could as well be used for non-composite/isotropic base materials. It may be noted that corrosion of metal pipes has become a key issue especially for marine applications like offshore oil rigs. High-performance composite pipes have been successfully utilized for this purpose.

Fig. 8.2 A plate with bonded smart layer for SHM

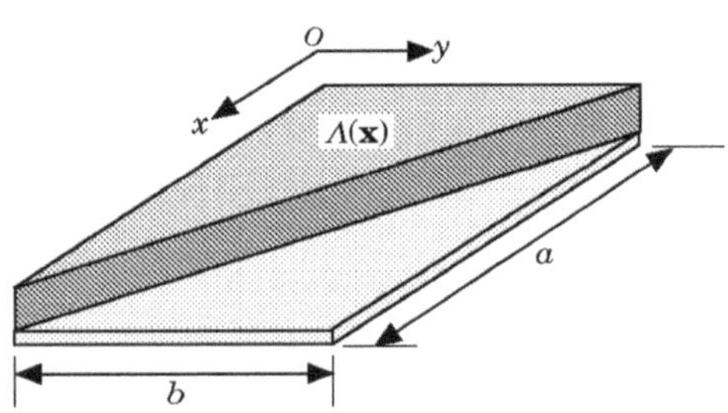

8.3.1 SHM for Actuation of Composite Beams

Composite Beam is a very common structural element used for application of SHM. Modeling of such beam is generally carried out using classical beam theory based on Euler–Bernoulli model. The reliability of the model can be validated with experimentally observed strain profile of a beam under combined stretching and bending. It is observed, that for a large range of slenderness ratio (ratio of beam span to thickness) this type of model predicts fairly accurate variation of strain across the thickness of the beam. Hence, in our discussion, in order to bring out the interaction mechanism of smart and host layer, the Euler–Bernoulli model is first taken up. The modeling exercise is divided into two sections. In the first section, symmetrically laminated beams are considered in which the coupling between in-plane and out–of-plane deformation is absent and hence the effect of smart layer addition on only the governing transverse equation of motion can be observed. In the next section, however, a more generalized case of unsymmetrical laminated beams are considered to discuss the coupling of in-plane and out-of-plane motion.

8.4 Distributed Vibration Control of Symmetrically Laminated Composite Beam

Figure 8.3 shows a typical multilayered symmetric composite beam with embedded smart layer. It is assumed that bonding is perfect between composite and smart layer. The thickness of the bonding layer is also neglected. The beam consists of n ply layers in which the m-th and $(n - m + 1)$-th layers are symmetrically placed with respect to the geometric mid-plane and are smart layers. These layers are made of either magnetostrictive or piezoelectric material. Fiber direction in the other composite layers could be arbitrarily specified except the condition that they are to be symmetric with respect to the mid-plane.

In this model for vibration suppression, spatially distributed actuation is considered. In the case of magnetostrictive actuation, such a situation can be realized by driving the required current through a series of closely placed, well-insulated coils along the span. Similarly, for piezoelectric actuation, different voltages may be

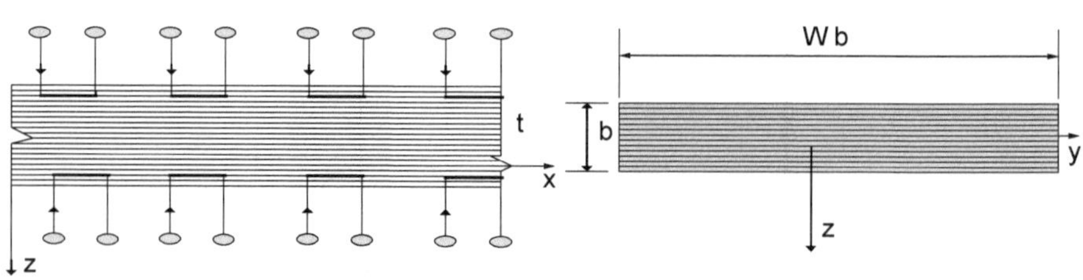

Fig. 8.3 Section of a laminated composite beam with embedded smart layer

applied independently to closely place well-insulated piezo-plates along the span. For closed-loop control, proportional velocity feedback control algorithm is employed as this kind of control law is known to provide stability as well as active damping. It is also assumed that the sensing is distributed and continuous in space and time. The beam is considered to be slender, so that the effect of rotary inertia and shear deformation can be neglected. The single layer theory of composite beam is used for analysis. The flexural motion of the beam in *XOZ* plane can be modeled by assuming the in-plane and out-of-plane displacement fields as

$$\begin{aligned} U(x,z,t) &= -zw_{,x}(x,t); \\ W(x,z,t) &= w(x,t) \end{aligned} \tag{8.4.1}$$

($_{,x}$) denotes differentiation with respect to x. The total strain in any (i-th) layer of the beam consists of two parts; the elastic strain ε_{x_s} which can be expressed by the following constitutive relationship:

$$\varepsilon^i_{x_s} = S\sigma^i_{x_s} \tag{8.4.2}$$

and the active strain ε_{x_a} which is present only in the active layer. Thus, the expression for strain in any layer can be written in generic form as:

$$\varepsilon^i_x = S\sigma^i_{x_s} + \delta_{ia}\varepsilon^i_{x_a} \tag{8.4.3}$$

where δ_{ia} is the Kronecker's delta. Its value is 1 when $i = a$, (i.e., for active layer) and zero for other layers. Also, from the constitutive relationship, for isotropic active layer, the compliance modulus $S = \frac{1}{E_a}$ and for orthotropic composite layers with fiber angle θ, the same may be expressed as:

$$S^i = (1/\bar{Q}^i_{11}) \tag{8.4.4}$$

$$\begin{aligned} \bar{Q}^i_{11} &= Q^i_{11}\cos^4\theta_i + 2(Q^i_{12} + 2Q^i_{66})\cos^2\theta_i\sin^2\theta_i + Q^i_{22}\sin^4\theta_i \\ Q^i_{11} &= \frac{E^i_{11}}{1 - v^i_{12}}v^i_{21},\ Q^i_{22}\frac{E^i_{22}}{1 - v^i_{12}v^i_{21}} \\ Q^i_{12} &= Q^i_{21} = \frac{v^i_{21}E^i_{11}}{1 - v^i_{12}v^i_{21}} \\ Q^i_{66} &= G_{12} \end{aligned}$$

where E^i_{11}, v^i_{21}, etc., are Elastic constants and Poisson's ratio, respectively. Following Euler–Bernoulli model, the total strain at any layer may also be expressed as

$$\varepsilon_x^i = -zw_{,xx}\,. \tag{8.4.5}$$

If we assume that the active strain is proportional to the transverse velocity of the beam, that is, $\varepsilon_x = C\dot{w}$, where the symbol '.' denotes differentiation with respect to time, the constant of proportionality C varies for different smart materials and control gain (f). The reason behind such assumption is that the active layer remains functional when the composite beam is vibrating at a high transverse velocity. This includes typical cases like small amplitude of displacement at a high frequency and large amplitude of displacement at a low frequency, both of which may cause failure of the structure. In the case of magnetostrictive material, $C = -d_m k_1 f$, where d_m is the magneto-mechanical constant and the coil constant k_1 is given by

$$k_1 = \frac{n_c}{\sqrt{w_c^2 + 4r_c^2}} \tag{8.4.6}$$

where n_c is the total number of coil turns, r_c is the effective length of magnetizing coil, and w_c is the effective width of magnetizing coil.

Thus, stress at any layer may now be written as

$$\sigma_x^i = \bar{Q}_{11}^i\left[\left(-zw_{,xx}\right) - C\left(\dot{w}\right)\right] \tag{8.4.7}$$

Using the Hamilton's principle, the governing equations of motion may be derived as

$$D_{11}w_{,xxxx} + \beta_{11}\dot{w}_{,xx} + m\ddot{w} = P(x,t) \tag{8.4.8}$$

with the following boundary conditions:

(i) either $w = 0$ or $D_{11}w_{,xxxx} + \beta_{11}\dot{w}_{,x} = P_0$;
(ii) either $w_{,x} = 0$, or $D_{11}w_{,xx} + \beta\dot{w} = M_0$;

Initial conditions in time will also be prescribed along with.

Here,

$$D_{11} = \sum_{i=1}^{n}\int_{z_1}^{z_{1+i}} \overline{\bar{Q}_{11}^i}z^2\mathrm{d}z$$

$$\beta_{11} = \sum_{i=1}^{n}\int_{z_k}^{z_{k+1}} S^{-1}C\delta_{ia}z\mathrm{d}z$$

$$m = \sum_{i=1}^{n}\int_{z_i}^{z_{i+1}} \rho\mathrm{d}z$$

$P(x, t)$ is the generalized distributed load on the beam, and P_0 and M_0 are concentrated load and moment on the beam applied at $x = x_0$.

The higher order terms in mass viz $\int_z \rho z\mathrm{d}z$ and $\int_z \rho z^2\mathrm{d}z$ are neglected. It is to be noted that for distributed control, there is an interesting difference in forming the equation of motion of the system. In the usual expression of Hamiltonian, the actuation of the structure is considered as an outside effect and the work done due to actuation is considered separately, whereas in this approach, the effect of actuator is considered implicitly in the evaluation of elastic strain (Eqs. 8.4.3 and 8.4.5). It is assumed that the total strain in the beam is governed by the displacement assumption corresponding to the beam model and the actual elastic strain is the difference between the total strain and the active strain. The advantage of this approach is that the control force comes out as an integral part of the model unlike a forcing function in the conventional approach. Let us exemplify the procedure to find out the response of a continuous system under distributed control.

Example 8.3.1 Consider a freely vibrating simply supported beam of length L, subjected to distributed control by active smart layer. There are two specified initial conditions of velocity 0 and 1 unit, respectively. Find out the responses of the system and compare it with uncontrolled vibration for a control gain of unity.

Solution Let us consider w, the transverse deflection as the product of the following spatial and temporal functions—

$$w(x,t) = T_1(t)\sin\frac{k\pi x}{L} \tag{8.4.9}$$

When, after variable separation, the free vibration governing equation based on (8.4.7) reduces to

$$DT_1 + \beta \dot{T}_1 + m\ddot{T}_1 = 0 \tag{8.4.10}$$

where

$$D = \frac{k^4\pi^4}{L^4}D_{11}$$
$$\beta = \frac{k^2\pi^2}{L^2}\beta_{11}$$

Considering $T_1 = X_I e^{\lambda t}$, where $\lambda = -\alpha \pm i\omega_d$, a complex eigen-value problem gets generated from Eq. (8.4.10). The damped natural frequency, ω_d and the damping constant, α of the system may be evaluated as:

$$\omega_d = \omega_n\sqrt{1-\varepsilon^2}, \quad \varsigma = \frac{c}{2m\omega_n} \tag{8.4.11}$$

$$\omega_n^2 = \frac{D}{m}, \quad c = \beta * f$$

$$\alpha = \frac{c}{2m} \tag{8.4.12}$$

Now, depending on the initial conditions in time, two different cases could be considered.

Case A: The solution for the case with initial velocity specified as $\dot{T}_1 = 0$;
Displacement:

$$w(x,t) = \frac{1}{\omega_d} \mathrm{e}^{\alpha t} \sin \omega_d t \sin \frac{\pi x}{L}$$

Actuation Effort (Voltage or Current):

$$E(t) = \frac{f}{\omega_d} \frac{\mathrm{d}}{\mathrm{d}t} [\mathrm{e}^{-\alpha t} \sin \omega_d t] \sin \frac{\pi x}{L}$$

Actuation Stress:

$$\sigma_a = \frac{C}{\omega_d} \frac{\mathrm{d}}{\mathrm{d}t} [\mathrm{e}^{-\alpha t} \sin \omega_d t] \sin \frac{\pi x}{L}$$

Case B: Similarly, for the case of initial velocity specified as $\dot{T}_1 = 1$, you may find

Displacement:

$$w(x,t) = \frac{\omega_d}{\sqrt{\alpha^2 + \omega_d^2}} \mathrm{e}^{-\alpha t} \left[\sin \left[\omega_d t - \tan^{-1} \left(\frac{-\omega_d}{\alpha} \right) \right] \right] \sin \frac{\pi x}{L}$$

Actuation Effort (Voltage or Current):

$$E(t) = \frac{f \omega_d}{\sqrt{\alpha^2 + \omega_d^2}} \frac{\mathrm{d}}{\mathrm{d}t} \left[\mathrm{e}^{-\alpha t} \sin \left(\omega_d t - \tan^{-1} \left(-\frac{\omega_d}{\alpha} \right) \right) \right] \sin \frac{\pi x}{L}$$

Actuation Stress:

$$\sigma_a = \frac{C \omega_d}{\sqrt{\alpha^2 + \omega_d^2}} \frac{\mathrm{d}}{\mathrm{d}t} \left[\mathrm{e}^{-\alpha t} \sin \left(\omega_d t - \tan^{-1} \left(-\frac{\omega_d}{\alpha} \right) \right) \right] \sin \frac{\pi x}{L}$$

The spatial variation for the actuation stress and the applied voltage is the same as that of deflection. However, the variation with respect to time is different for deflection and application of electric/magnetic field. The present solution is although obtained for the first mode of vibration of a simply supported beam, it should be noted that the controller is not designed individually for each mode. It is shown through the solution of transverse deflection w that the presence of constant gain velocity feedback can suppress any mode of vibration and hence the overall vibration of the beam. Thus, the problem of 'spillover' does not arise here. The general solution of deflection shows that the damping effect of smart actuator is implicitly incorporated through the term α, since

$$\alpha = \frac{\beta f}{2m}.$$

By controlling the control gain and thereby the term β, one can achieve the required damping. The presence of smart layer also affects the natural frequency of the beam through both damping coefficient ξ and material property constant D_n. In the case of asymmetric beam, additional dependency of frequency and damping on in-plane inertia arises which will be dealt in the next section.

8.5 Asymmetric Composite Beam

In general, the ply layups in composite laminate may be designed as symmetric from the structural point of view. The effectiveness of the distributed induced strain actuation for a given structure will depend on the positioning of the active layer which may not be symmetrically located with respect to geometrical neutral axis. Also, inclusion of a single active layer is often preferred to multiple layers as only one set of coils are sufficient for such actuation and hence structural integrity remains less disturbed. However, from modeling point of view, this creates additional problems since the in-plane and out-of-plane motion gets coupled in such cases and we will consider such cases in this section.

A laminated beam with n layers containing a single layer of magnetostrictive or piezoelectric material is considered in the present analysis. The result can be easily generalized for multiple arbitrarily placed active layers. The layup details are similar to Fig. 8.3 except that the mth layer is passive in this case instead of being smart layer. The presence of active layer on one side of the beam introduces asymmetry and as a consequence, the in-plane displacement is nonzero along x-axis at $z = 0$. Including this, the displacement field may be modified from Eq. (8.4.1) as,

$$U(x,z,t) = u(x,t) - zw_{,x} \tag{8.5.1}$$

$$W(x,t) = w(x,t) \tag{8.5.2}$$

Consequently, the expression for strain becomes

$$\varepsilon_x = u_{,x} - z w_{,xx} \tag{8.5.3}$$

The constitutive relations and the expression for H remain the same as in the previous case (Eq. 8.4.6). The governing equations of motion and the boundary conditions may be derived using Hamilton's Principle as,

$$\begin{aligned} A_{11}u_{,xx} - B_{11}w_{,xxx} - F_1 c\dot{w}_{,x} - m\ddot{u} &= 0 \\ -B_{11}u_{,xxx} + D_{11}w_{,xxx} + F_2 c\dot{w}_{,xx} + m\ddot{w} &= 0 \end{aligned} \tag{8.5.4}$$

where

$$A_{11}, B_{11}, D_{11} = \sum_{i=1}^{n} \int_{z_i}^{z_{i+1}} \bar{Q}_{11}^{(i)}(1, z, z^2)\,\mathrm{d}z$$

$$F_1, F_2 = \int_{z_m}^{z_{m+1}} C(1, z)\delta_{ia}\,\mathrm{d}z$$

$$m = \sum_{i=1}^{n} \int_{z_i}^{z_{i+1}} p^{(i)}\,\mathrm{d}z$$

The boundary conditions at $x = 0$ and L are

$$\begin{aligned} &\text{either } w = 0 \text{ or } B_{11}u_{,xx} - D_{11}w_{,xxx} - F_2 C\dot{w}_{,x} = 0 \\ &\text{either } w = 0 \text{ or } B_{11}u_{,x} - D_{11}w_{,xx} - F_2 C\dot{w} = 0 \\ &\text{either } w = 0 \text{ or } A_{11}u_{,x} - B_{11}w_{,xx} - F_2 C\dot{w} = 0 \end{aligned} \tag{8.5.5}$$

Let us further illustrate the usefulness of Eq. (8.5.4), in obtaining the response of an asymmetric simply supported beam.

Example 8.4.1 Consider the vibration control of a simply supported asymmetric composite beam. Find out the response of the system from the governing equation of motion.

Solution At simply supported end, the boundary conditions are

$$\begin{aligned} &w = 0 \\ &B_{11}u_{,x} - D_{11}w_{,xx} - F_2 c\dot{w} = 0 \\ &A_{11}u_{,x} - B_{11}w_{,xx} - F_2 C\dot{w} = 0 \end{aligned} \tag{8.5.6}$$

They can be satisfied by making $w = 0, w_{,xx} = 0$ and $u_x = 0$ in the governing equation. To satisfy these conditions, u and w may be chosen as,

$$u(x,t) = T_1(t)\sin\frac{n\pi x}{L} \tag{8.5.7}$$

$$w(x,t) = T_2(t)\cos\frac{n\pi x}{L} \tag{8.5.8}$$

Substituting Eqs. (8.5.7) and (8.5.8) in Eq. (8.5.6), we get

$$AT_2 - BT_1 - F_1\dot{T}_1 + m\ddot{T}_2 = 0 \tag{8.5.9}$$

$$-BT_2 + DT_1 - F_2\dot{T}_1 + m\ddot{T}_1 = 0 \tag{8.5.10}$$

with

$$A = \frac{n^2\pi^2}{L^2}A_{11}, \quad B = \frac{n^3\pi^3}{L^3}B_{11}, \quad D = \frac{n^4\pi^4}{L^4}D_{11}$$

$$F_1^* = \frac{n\pi}{L}C, \quad F_2^* = \frac{n^2\pi^2}{L^2}C \tag{8.5.11}$$

The gain C to be chosen, (as in the previous case) is a constant as $C = -d_{31}$ (for piezoelectric material) or $C = -d_m k_1 f$ (for magnetostrictive material).

Considering $(T_1T_2) = (X_1, X_2)e^{\lambda t}$ and substituting in Eqs. (8.5.9) and (8.5.10), for nontrivial solution, one gets.

$$\begin{vmatrix} A + m\lambda^2 & -(B + F_1^*\lambda) \\ -B & D + F_2^*\lambda + m\lambda^2 \end{vmatrix} \tag{8.5.12}$$

Equation (8.5.12) provides two sets of complex eigen-values, one representing primarily transverse motion and the other axial motion. The frequencies associated with the axial motion are very high compared to frequency of primarily transverse motion. Hence, we are concerned with primarily transverse motion and it is represented by the lowest eigen-values of Eq. (8.5.12) in the form $\lambda = -a \pm i\omega_d$ considering initial conditions as $T_1 = 0$, $\dot{T}_1 = 0, \dot{T}_2 = 1$ at $t = 0$, the expressions for w and u may be obtained as,

$$w = \frac{1}{\omega_d}e^{-\alpha t}\sin\omega_d t \sin\frac{n\pi x}{L} \tag{8.5.13}$$

$$u = \frac{\phi}{\omega_d}e^{-\alpha t}\sin\omega_d t \cos\frac{n\pi x}{L} \tag{8.5.14}$$

$$\phi = \left|\frac{-B - F_1^*\lambda}{A + m\lambda^2}\right| \tag{8.5.15}$$

For magnetostrictive material, the actuation stress is $\sigma_a = -E_m dH$, and the actuation current I in the magnetizing coil depends on the number of turns per unit length and coil radius. Assuming 2 cm coil radius and coil density of 10^4 turns per meter, the coil current may be obtained as,

$$I(t) = \frac{-f}{9992}\omega_d \sin\frac{n\pi x}{L}\frac{\mathrm{d}}{\mathrm{d}t}\left(\mathrm{e}^{-\alpha t}\sin\omega_d t\right) \tag{8.5.16}$$

Similarly, the actuation stress σ_a is obtained as

$$\sigma_a(t) = \frac{E_m df}{\omega_d}\sin\frac{n\pi x}{L}\frac{\mathrm{d}}{\mathrm{d}t}\left(\mathrm{e}^{-\alpha t}\sin\omega_d t\right) \tag{8.5.17}$$

In the last two examples, we have explained how SHM can be successfully applied for controlling vibration of structures like composite plates.

Another application of SHM is in the field of sensing damages of composite structure.

8.6 SHM for Sensing Damages of Composite Structure

The performance of laminated composites may get degraded due to defects like fiber breakage, matrix cracking, de-bonding between fibers and matrix, and delamination or interlayer cracks. These defects can significantly affect the mechanical properties of structures and may eventually lead to failure. What follows is a brief description of a novel method of detecting damages in laminated composites using smart sensors. To begin with, for the sake of brevity, we will discuss the essential macro-mechanical relationships.

For simplicity, only composite laminate with unidirectional fibers is considered and the thickness of each layer is considered to be uniform. The elastic moduli of the composites are derived from the moduli of the fiber and of the matrix by using the standard micro-mechanical formulations:

$$\begin{aligned}
E_1 &= E_f V_f + E_m V_m \\
\nu_{12} &= \nu_f V_f + \nu_m V_m \\
\frac{M}{M_m} &= \frac{1+\zeta\eta V_f}{1-\eta V_f} \\
\eta &= \frac{(M_f/M_m)-1}{(M_f/M_m)+\zeta}
\end{aligned} \tag{8.6.1}$$

where E: Elastic Modulus; G: Shear Modulus; ν: Poisson's Ratio; M: Composite Elastic Modulus E_2, G_{12}, or ν_{12}; M_f: Corresponding fiber modulus E_f, G_f or ν_{12}; M_m: Corresponding Matrix Modulus E_m, G_m, ν_m; V_f: Fiber Volume Fraction; $\zeta = 2$ for E_2 and 1 for G_{12}.

The aforementioned equations provide the equivalent mechanical properties of an individual lamina. However, for strength and loading analysis, it is generally more convenient to express the combined properties of all the lamina, i.e., effective laminate engineering constants. The following equations establish the relationship between the stress resultants and strains in the system.

$$\begin{Bmatrix} N_x \\ N_y \\ N_{xy} \\ M_x \\ M_y \\ M_{xy} \end{Bmatrix} = \begin{pmatrix} A_{11} & A_{12} & A_{16} & B_{11} & B_{12} & B_{16} \\ A_{12} & A_{22} & A_{26} & B_{12} & B_{22} & B_{26} \\ A_{16} & A_{26} & A_{66} & B_{16} & B_{26} & B_{66} \\ B_{11} & B_{12} & B_{16} & D_{11} & D_{12} & D_{16} \\ B_{12} & B_{22} & B_{26} & D_{12} & D_{22} & D_{26} \\ B_{16} & B_{26} & B_{66} & D_{16} & D_{26} & D_{66} \end{pmatrix} \begin{Bmatrix} \varepsilon_x^0 \\ \varepsilon_y^0 \\ \gamma_{xy}^0 \\ k_x \\ k_y \\ k_{xy} \end{Bmatrix} \tag{8.6.2}$$

where

N_x Stress resultant along x direction per unit width (N m^{-1})
N_y Stress resultant along y direction per unit width (N m^{-1})
N_{xy} Shear stress resultant force per unit width (N m^{-1})
M_x Moment resultant along x direction per unit width (N)
M_y Moment resultant along y direction per unit width (N)
M_{xy} Torsional moment resultant per unit width (N)
h_k Height of the *k*th layer

In the above formulation, the sub-matrices *A*, *B*, and *D* are as follows:

$$\begin{aligned} [A] &= \sum_{k=1}^{n} \left[\overline{Q}_{ij}\right]_k (h_k - h_{k-1}) \\ [B] &= \frac{1}{2}\sum_{k=1}^{n} \left[\overline{Q}_{ij}\right]_k \left(h_k^2 - h_{k-1}^2\right) \\ [D] &= \frac{1}{3}\sum_{k=1}^{n} \left[\overline{Q}_{ij}\right]_k \left(h_k^3 - h_{k-1}^3\right) \end{aligned} \tag{8.6.3}$$

A, *B*, and *D* are in-plane stiffness, in- and out-of-plane coupling, and out-of-plane coupling matrices, respectively. Here,

$$T = \begin{bmatrix} \cos^2\theta & \sin^2\theta & 2\sin\theta\cos\theta \\ \cos^2\theta & \sin^2\theta & 2\sin\theta\cos\theta \\ -\sin\theta\cos\theta & \sin\theta\cos\theta & \cos^2\theta - \sin^2\theta \end{bmatrix} \tag{8.6.4}$$

θ Fiber inclination angle from the *x*-axis

where $\overline{Q}$ is the stiffness matrix of a lamina with ply angle θ and Q is the stiffness matrix of a lamina with 0° ply angle. With this background of macro-mechanics, we will first consider the constitutive relationship of piezoelectric materials which we have already discussed in the last chapter. The constitutive relationship of the piezoelectric material may be written as:

$$\begin{aligned} D &= d\sigma + \varepsilon^{*\sigma} E \\ \varepsilon &= S^E \sigma + dE \end{aligned} \tag{8.6.5}$$

where D is the electric field displacement, d is the piezoelectric charge constant, $\varepsilon^{*\sigma}$ is the permittivity matrix at constant stress field, E is the electric field intensity, ε is strain, and s^E is the compliance matrix.

On application of external electric field, the state of stress of the composite can be represented by Fig. 8.4. With respect to the piezo-layer, the composite gets divided into two sub-laminates.

The upper laminate is having p_1 number of layers and a resultant tensile stress N_1; lower laminate is having p_2 number of layers and a resultant tensile stress N_2. Since, it is a general case of loading, bending is also expected, which is represented as M. Further, the following relations hold true:

$$\begin{aligned} \{N_1\} &= [A_1]\{\varepsilon_p\} \\ \{N_2\} &= [A_2]\{\varepsilon_p\} \\ \{N\} &= \{N_1\} + \{N_2\} \end{aligned} \tag{8.6.6}$$

Here, A_1 and A_2 are the in-place stiffness matrices of the composite layers above and below the piezo, respectively. N_1 and N_2 are the force/width of the piezo, and ε_p is the strain in the piezo-patch given by:

$$\{\varepsilon_p\} = \{\varepsilon_0\} + z_p\{k\} \tag{8.6.7}$$

$\{\varepsilon_0\}$ is the mid-place strain, z_p is the distance of the piezo from the mid-plane, and κ is the curvature of the composite. The moment balance yields—

$$\{M\} = \{N_1\}a + \{N_2\}b - \{N\}c \tag{8.6.8}$$

Fig. 8.4 A composite laminate with embedded sensing layer

where the distances a, b, and c are illustrated in Fig. 8.4. Also, available is the force balance equation on the piezoelectric layer as:

$$\{N\} = \{\sigma_p\}h_p \tag{8.6.9}$$

where σ_p and h_p are the stress on the piezo-layer and its distance from the top, respectively. All the above equations can be solved simultaneously to obtain the electric field E as follows:

$$\{\varepsilon\} = ([d][\alpha_3][d] + [\mu])^{-1}\{D\}$$

where

$$\begin{aligned} [\alpha_3] &= \left[\overline{A_{12}}h_p - [s] + y_p h_p[\alpha_2]\right]^{-1} \\ [\alpha_2] &= [D]^{-1}\left(\alpha_1 - [B]\overline{A_{12}}\right)\{\sigma_p\}h_p \end{aligned} \tag{8.6.10}$$

$$[\alpha_1] = A_1\overline{A_{12}}\alpha + A_2\overline{A_{12}}c - Ib, \overline{A_{12}} = [A_1 + A_2]^{-1}$$

Here, I is an identity matrix of size 3 × 3. Furthermore, following the constitutive relationship, stress on the piezoelectric layer can be expressed as,

$$\{\sigma_p\} = [\alpha_3][d]\{E\} \tag{8.6.11}$$

Here, E is the electric field generated due to the self-sensing capabilities of the piezo-patch. To implement such self-sensing feature, the circuit shown in Fig. 8.5 may be used.

The electric field and the finally obtained voltage V_p are related as follows:

$$V_p = ([d][\alpha_3][d] + [\mu])^{-1}\{D\}t_p \tag{8.6.12}$$

where t_p is the thickness of the piezo-patch.

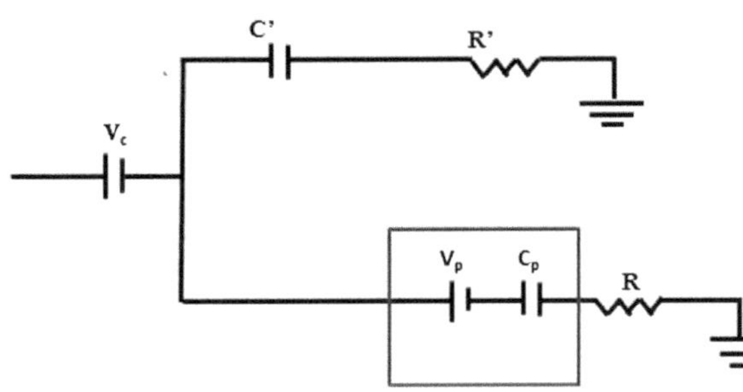

Fig. 8.5 A self-sensing actuation circuit for piezoelectric material. The red block shows the equivalent electrical model of the piezo-actuator

8.7 Damage Identification Criteria for SHM

In the last section, we have shown that smart layers embedded in laminated composite can be effectively used for sensing the state of stress and thereby the change of axial/bending stiffness of the composite laminate due to damage. In order to localize the damage based on dynamic response of laminated composite, several damage criteria are used. In this section, we will deal with some of these criteria.

It is generally observed that a crack or localized damage could reduce stiffness and increase damping in the structure. Reduction in stiffness is generally associated with decrease in the natural frequencies of the structure and modification of the modes of vibration of the structure. Simple dynamic response-based damage identification techniques are hence based on change in natural frequency or damping. Even though these techniques can be used effectively to determine the presence of crack or damage in a structure, but, they are quite ineffective in predicting the location of the damage. This is because cracks at two different locations associated with certain crack lengths may cause same amount of frequency change in the structure. More accurate methods that can both detect and locate damages are based on the study of a number of modal parameters which will be elucidated in the following section.

Modal methods of damage identification are based on the fact that the modal parameters such as mode shape, modal damping, and natural frequency are functions of the physical parameters like mass, stiffness, and damping of the structure. Hence, it can be assumed that the presence of damage may lead to some changes in some of the modal properties of the structure. Damage detection procedures also inspect the change in parameters like modal curvature that takes place when the structure is damaged. The modal parameters that are generally used to detect structural damage are frequency response functions (FRF), natural frequencies, mode shapes; modal curvature shapes (MCS), modal flexibility, and modal strain energy. It may be stated that not all these parameters are equally sensitive to predict the presence of damage in a structure. For example, changes in the mass of the structure and ambient temperature may generate uncertainties in the measured frequency changes [4]. The difficulty of damage detection based on change in frequency can be overcome to some extent by using changes in the curvature mode shapes of the given structure. In general, mode shapes are more sensitive compared to natural frequency for detecting local damage. The curvature mode shape change due to damage has a local effect in nature, so it can be used to locate damage properly. The curvature mode shape methods have good potential to identify damage type such as delamination.

The basic ideas of structural health monitoring are well suitable in the laboratory. But how one can implement the SHM methodology to the real structures? Various methods of SHM are available in this regard. These methods are generally classified as model-based methods and signal-based methods and are elucidated in Fig. 8.6.

In signal-based methods, from the measurement data, some characteristic features, such as means, variances, maximum/minimum values, spectral information,

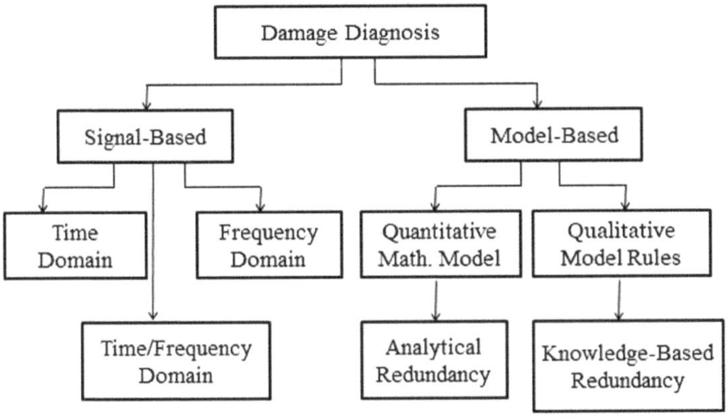

Fig. 8.6 Various methods of damage diagnosis

are extracted by pure signal analysis. As stated in Axiom II, we compare the actual values with the reference values which are derived from the undamaged state of the monitored system. This is generally carried out in time or frequency domain. However, the combined time–frequency analysis, also called wavelet analysis, has become another important tool in the data analysis. The signal analysis methods are generally useful for damage detection problems only.

The model-based techniques are based on the qualitative and quantitative models of the monitored system. The use of these models to interpret the measurement data opens up the horizon for a much wider field of application in damage diagnosis. The quantitative models comprise all types of mathematical models. For example, structural models describe the static or dynamic behavior of the mechanical system, input–output state space models, and integrated electromechanical (E/M) impedance method with a finite element model. In qualitative models, some knowledge rules are developed and from the measured data some characteristic features are extracted, and the developed rules are used to decide the type of problem. These knowledge rules are supported by decision fusion. From the data acquisition to the decision making, damage diagnosis is combined into a Data to Decision process (D2D) [3]. The basic model of a D2D process also known as the Waterfall model is explained in Fig. 8.7.

As apparent from Fig. 8.7, the first stage of D2D process gets initiated with the sensing of the signal. The raw data are further processed to filter out noise and reduce the data dimension by suitably transforming the data into frequency domain. Once we convert the entire data into a suitable de-noised form, we carry out the feature extraction process by marking the distinguished patterns/features and storing them in a feature vector. This can be carried out either using engineering judgment or statistical algorithms like principal component analysis. In the pattern processing phase, the feature vectors are studied to judge the type and extent of damage presence in the structure. These are usually carried out using three different techniques: novelty detection technique (allows unsupervised learning), damage

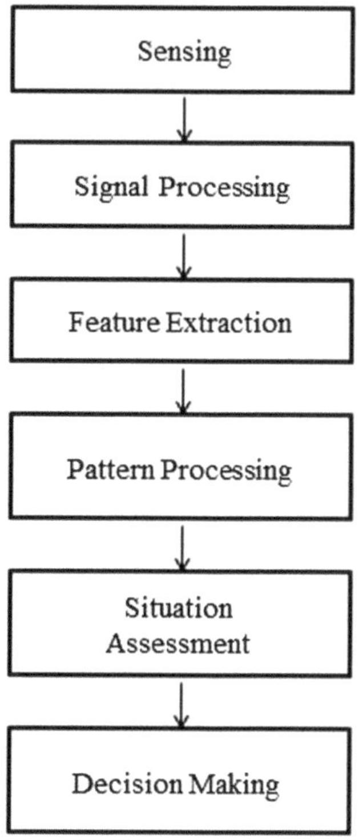

Fig. 8.7 Waterfall model of D2D

classification algorithms (discrete output), and regression analysis (continuous output). At the final two stages, decisions are arrived at (regarding the state of damage) based on the outcome of the pattern processing. The waterfall method, even though allows multiple sensor data to be fused at the stage of decision making; it is by and large an unidirectional model. Also, it does not contain any action item in the algorithm. Hence, Omnibus model as shown in Fig. 8.8 is often considered to be a better model of sensor data fusion.

Damage diagnosis problem can be solved both in supervised learning mode as well as unsupervised learning mode. Unsupervised learning mode does not involve the use of any particular damage class. Instead of learning an input–output mapping, the goal is to model the probability distribution of the input data or to discover clusters in the data. The novelty detection methods such as outlier analysis, kernel density methods, and auto-associative neural network are unsupervised learning methods [2]. In the supervised learning mode the knowledge rule is trained by a set of target values from the network outputs. The data may be obtained from modeling/ computation and experiment. For each set of input patterns which are the characteristic features extracted from the raw data, a desired network output is specified which is the characteristic of a particular damage class. By applying the knowledge rule to the defined input–output relations known as learning, the algorithm is able to recognize certain damage class when applied to measured data of the structure. There

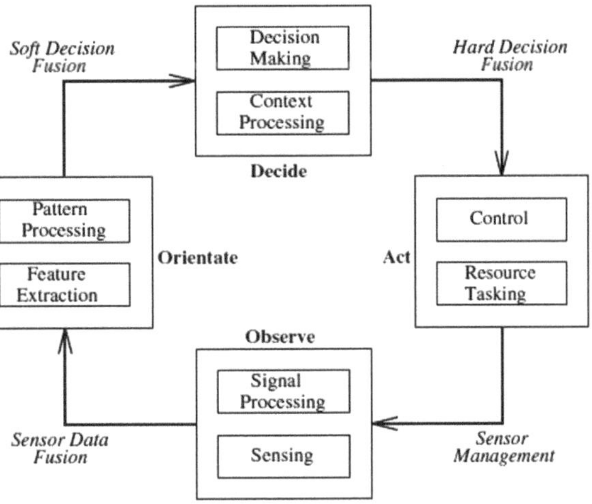

Fig. 8.8 Omnibus model of data fusion [3]

are many classification and regression techniques such as Linear Discriminant Analysis, Quadratic discriminant analysis, Nearest neighbor classifier, Kernel discriminant analysis, Support vector machines for further reading [2–4].

Ensemble methods in machine learning are the methods for combining the multiple models (base classifiers) each of which solve the same original task to obtain the better predictive performance that could be obtained from any of the single model. Ensemble methods are basically techniques for combining multiple base classifiers to produce strong classifier. These are supervised learning methods, because these methods can be trained by using training data and then used to make the predictions on unknown data. Every ensemble system has two main goals: first is to build an ensemble that is as diverse as possible and second is to combine the outputs of individual classifier. Various ensemble methods used are Bagging, Boosting, Rotation forest, etc.

(a) **Bagging**

Bagging is a short form of bootstrap aggregating in which different training data subset are randomly drawn with replacement from entire training data set. Each training data subset is used to train a different classifier of the same type. Decision obtained by each individual classifier is combined by using majority voting. The given object is classified on the basis of class chosen by most of the classifiers from the ensemble system.

(b) **Rotation Forest**

Rotation Forest is an ensemble method which trains L-base classifier independently, using a different set of extracted features for each base classifier. For building the classifier in rotation forest, bootstrap samples are taken from the training data set. On these samples, principal component analysis (PCA) is carried out separately and a new set of features are extracted by pooling all the principal components. The data are transformed linearly into the new feature space. The classifier is trained by using this extracted feature set. Different splits of the feature set will lead to different extracted features resulting in improvement in diversity by the bootstrap sampling.

So far we have carried out only abstract discussion on damage identification. In the following section, we will give some practical examples of feature selection from the dynamic response of a vibrating system.

8.8 Features of Various Damage Indices

In this section, we will present three different features that are used on the finite element models to identify the presence of damage and damage location. They are curvature mode shape (CMS), curvature damage factor (CDF), and damage index method (DIF), respectively [5–11].

Curvature Mode Shape (CMS): When damage is introduced in a structure, the bending stiffness at the location of the damage is reduced while at the same time the magnitude of the curvature modes increases. In the current study, CMS indicates the strain per unit of element per unit length. The absolute difference between the curvature modes of the healthy and damaged structures is defined as:

$$\Delta\varphi''_{ij} = \left|\{\varphi''_h\} - \{\varphi''_d\}\right|_{ij} \tag{8.8.1}$$

where φ''_{ij} is the ith curvature mode shape; h and d denote healthy and damaged conditions, respectively; and i and j denote the mode number and the measurement location, respectively. This is the highest in the region of the damage and negligibly small in the remaining region. This method examines each mode individually, and it is classified as a single-mode-based damage detection method. Results from this method can vary depending on the boundary conditions, damage locations, mode of interest, and sensitivity.

Curvature Damage Factor: The evaluation of curvature damage factor (CDF) method involves a similar procedure as the absolute difference method; however, the curvature damage factor is developed to consider all the modes of the vibrating system. The CDF is the summation of damage differences from each mode being evaluated.

$$\mathrm{CDF} = \frac{1}{N}\sum_{i=1}^{N}\left|\varphi''_{h,i} - \varphi''_{d,i}\right| \tag{8.8.2}$$

where CDF is the curvature damage factor at location i and N is the number of modes that will be examined. This method is considered to be more accurate than the absolute difference method because it eliminates the chances of insensitivity of a damage location associated with certain modes.

Damage Index Method: In comparison to CDF, the damage index method (DIM) allows for greater sensitivity. This method is also more complex as compared to the other two methods. Based on the differences of curvature mode ϕ'' at a certain location and CDF, the location of damage in the structure can be identified. However, closer to boundary, the fixed surface shows higher value of CDF and significant changes in curvature mode shapes of a composite plate. To overcome this problem, damage index method may be used effectively.

The damage index can be represented as:

$$\beta_{ij} = \frac{\left[\left\{\varphi''_{ij,d}\right\}^2 + \sum_{i=1}^{i_{\max}}\left\{\varphi''_{ij,d}\right\}^2\right] \times \sum_{i=1}^{i_{\max}}\left\{\varphi''_{ij,h}\right\}^2}{\left[\left\{\varphi''_{ij,h}\right\}^2 + \sum_{i=1}^{i_{\max}}\left\{\varphi''_{ij,h}\right\}^2\right] \times \sum_{i=1}^{i_{\max}}\left\{\varphi''_{ij,d}\right\}^2} \tag{8.8.3}$$

where ϕ''_{ij} is the ith curvature mode shape; h and d denote healthy and damaged laminate, respectively; i and j denote the mode number and the measurement location, respectively.

These indices are now evaluated for a typical case of a laminated composite plate. One of the most common yet complex cases of damage in composite laminate is called delamination. A composite plate is generally composed of number of ply layers bonded together. However, occasionally due to improper manufacturing, moisture absorption, etc., the ply layers may get separated from each other locally. This is known as delamination. Figure 8.9 shows different cases of delamination positions and the size of delamination considered in the analysis. The size of the composite plates chosen is 250 mm × 200 mm × 2.6 mm. For all these cases, the fiber-angle orientations considered are [0/0/0/0], [0/90/0/90] (also called cross-ply laminate), [90/90/90/90], respectively, and the thickness of the lamina is taken as 0.5 mm. A typical delamination of thickness same as for lamina is considered in the form of Teflon insert in all the cases.

Figures 8.10 and 8.11 show the mode shapes from modes 1–4 of a typical laminated composite plate.

Mode 1 is recognized as the first bending mode. Mode 2 is the first torsional mode. Mode 3 is a combination of bending and torsional modes. Mode 4 represents the second bending mode. Table 8.2 shows the fundamental frequencies of healthy plates as obtained from the finite element simulation for all the three types of plates without delamination. This can be used for reference.

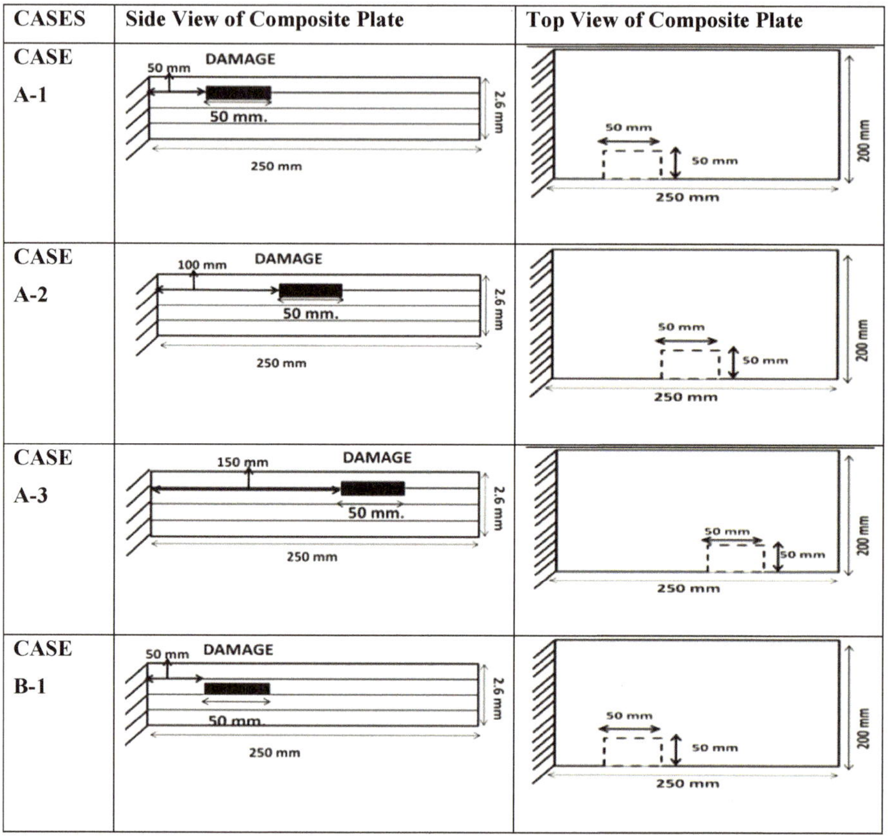

Fig. 8.9 Laminated composite plates with different positions of delamination

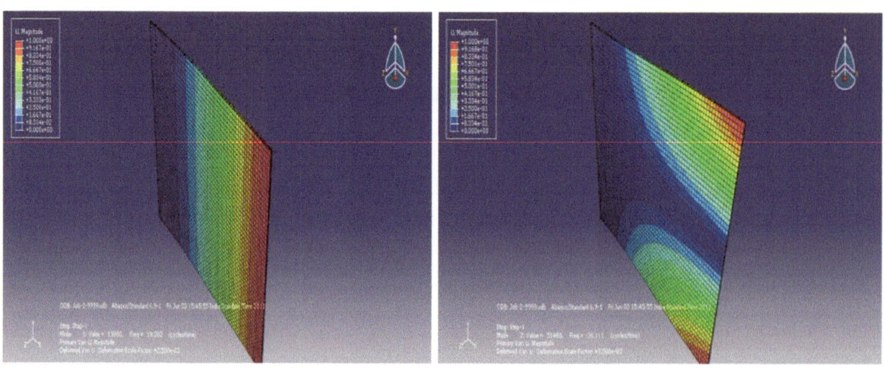

Fig. 8.10 Mode shapes of mode 1 and 2 of healthy composite plate

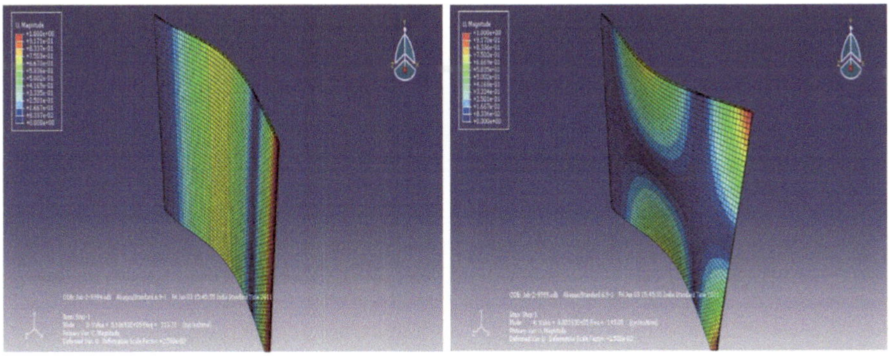

Fig. 8.11 Mode shapes of mode 3 and 4 of healthy composite plate

Table 8.2 Fundamental frequencies of the healthy plates

Mode No.	[0/0/0/0] (Hz)	[0/90/0/90] (Hz)	[90/90/90/90] (Hz)
MODE-1	7.38	18.2	26.88
MODE-2	29.18	36.11	47.67
MODE-3	95.57	113.73	98.83
MODE-4	129.24	143.08	167.85

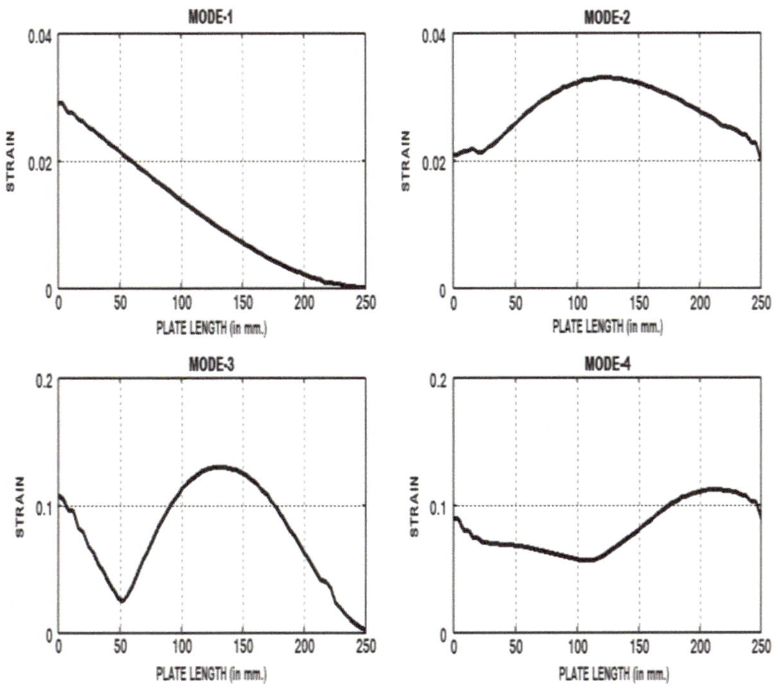

Fig. 8.12 First four CMS of healthy plate

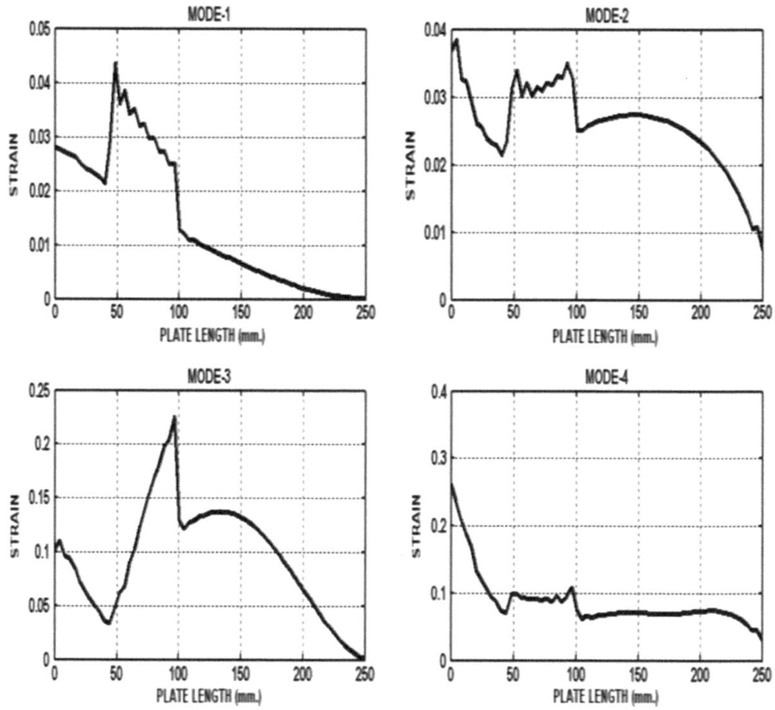

Fig. 8.13 First four CMS of case A-1 plate

All the four modes of the composite plate have been studied here with their curvature mode shapes (CMS). For calculating CMS of each case, the strain rate has been calculated for each element of the composite plate and the average across the width of the plate is taken. The element size has been taken as 0.004 mm. In this analysis, eight modes have been considered for each case of delamination. Delaminations are provided between first–second ply and second–third ply in all the cases. Figure 8.12 shows the CMS for a healthy [0/0/0/0] composite plate for the first four modes. The smooth continuity of the modal curves reflects the absence of any damage due to delamination. CMS of healthy plate shows some minor disturbance at the fixed end. These disturbances are generally present due to boundary conditions. Figure 8.13 shows the CMS of damaged composite plates for the same laminate for case A-1, considering the first four modes, respectively. CMS

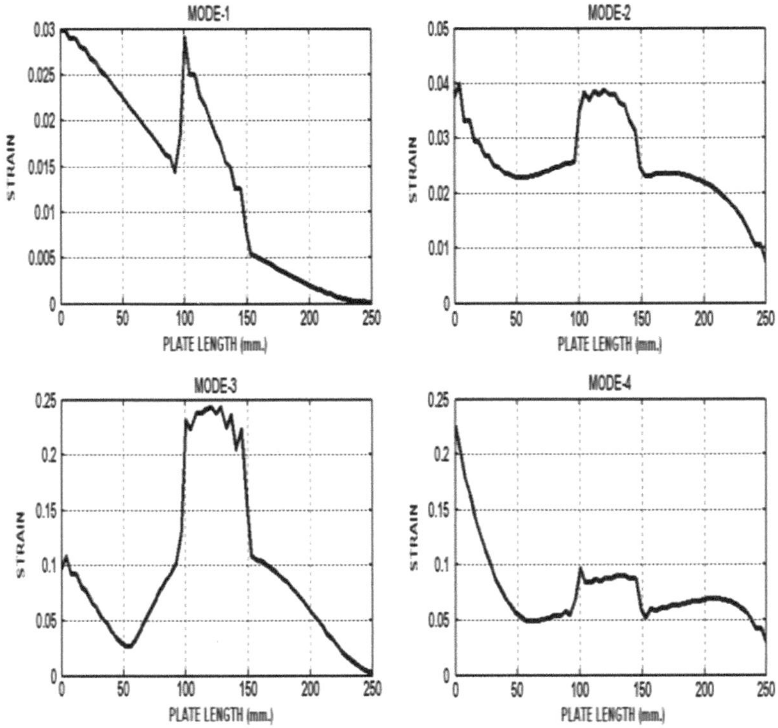

Fig. 8.14 CMS of case A-2 plate

corresponding to case A-1 clearly shows the presence of undulations between 50 and 100 mm along the plate length. These peaks are more prominent for Modes 1, 2, and 4 as the position of damage is more sensitive to these modes. Similar results are observed for Case A-2, A-3 and B-1 plates and are shown in Figs. 8.14, 8.15, and 8.16 respectively. Figures 8.17, 8.18, 8.19, and 8.20 show the CDF corresponding to the four cases of delamination in damaged plates, A-1, A-2, A-3, and B-1, respectively. Again, in all the cases except for case A-3, the CDF is successful in predicting the presence of delamination. The delamination is relatively far away from the boundary in case of Case A-3, where possibly the variation of modal strain is smaller. Figures 8.21, 8.22, 8.23, and 8.24 show the results of DIM-based analysis which is successful in predicting damage location in all the cases.

Thus, for the current feature selection, one may say that DI is more versatile in picking up all the damage cases and hence more preferable to the other two.

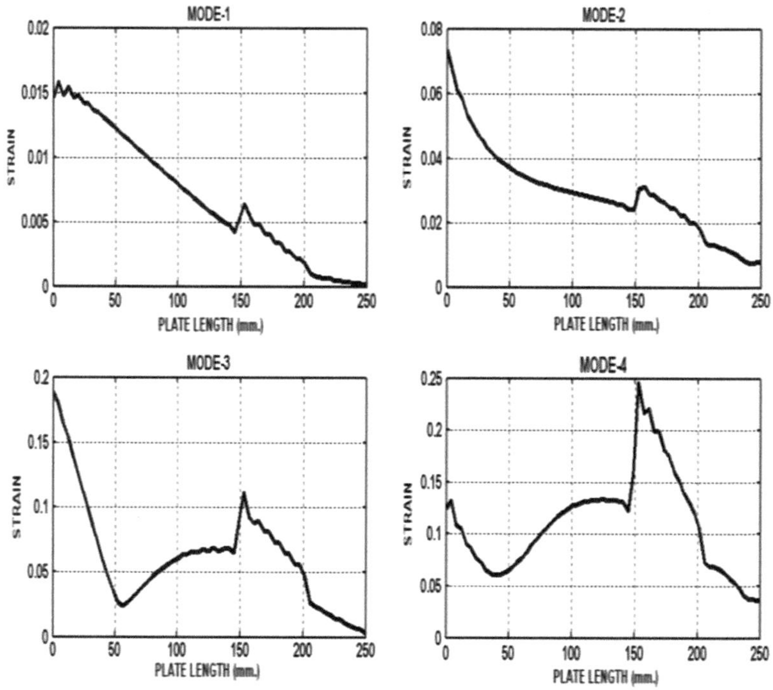

Fig. 8.15 CMS of case A-3 plate

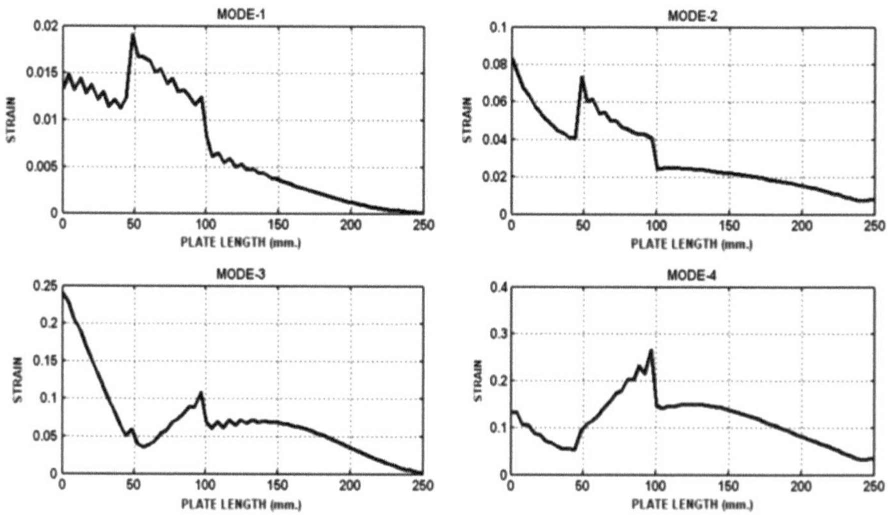

Fig. 8.16 CMS of case B-1 plate

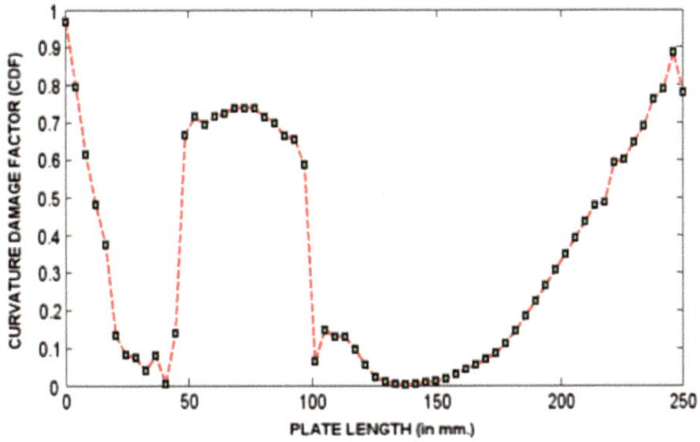

Fig. 8.17 CDF of CASE A-1 plate

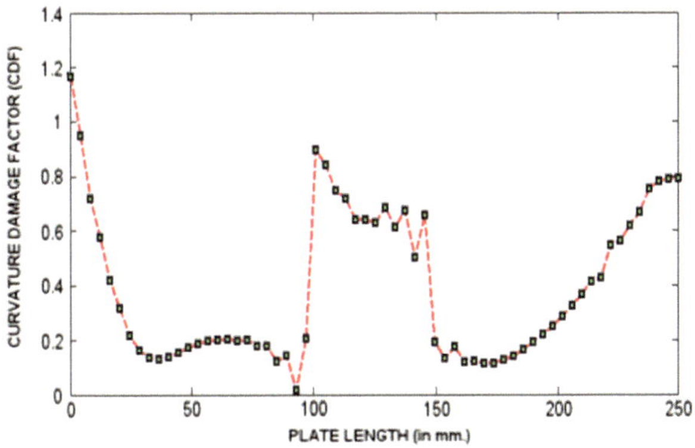

Fig. 8.18 CDF of CASE A-2 plate

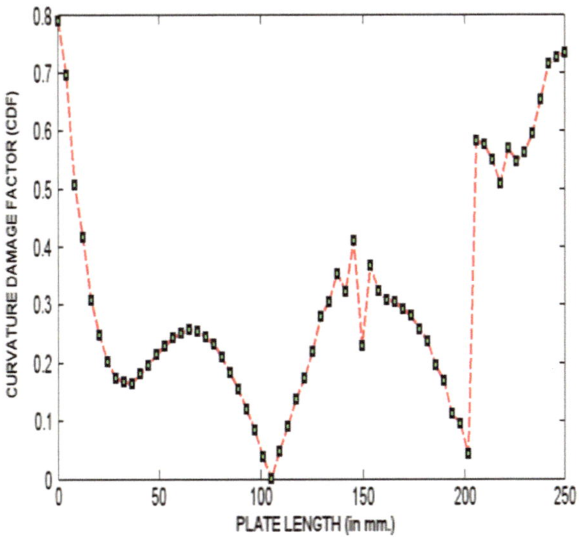

Fig. 8.19 CDF of CASE A-3 plate

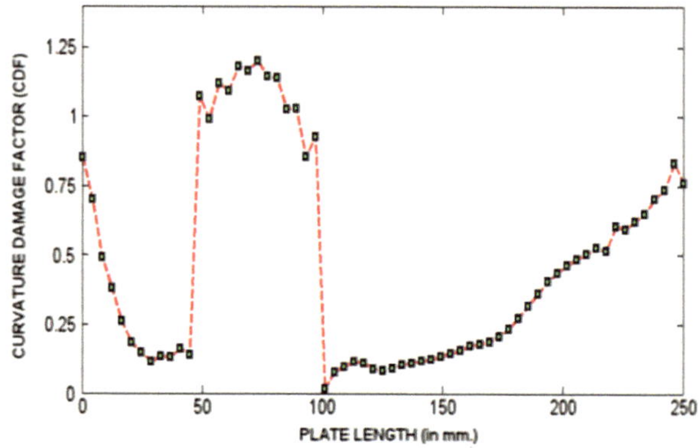

Fig. 8.20 CDF of CASE B-1 plate

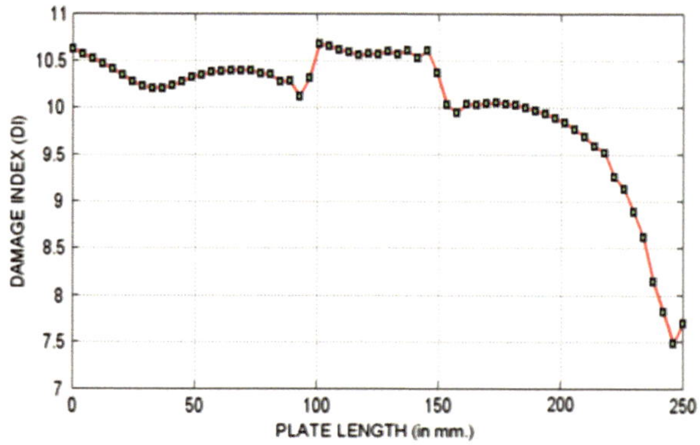

Fig. 8.21 DI of CASE A-1 plate

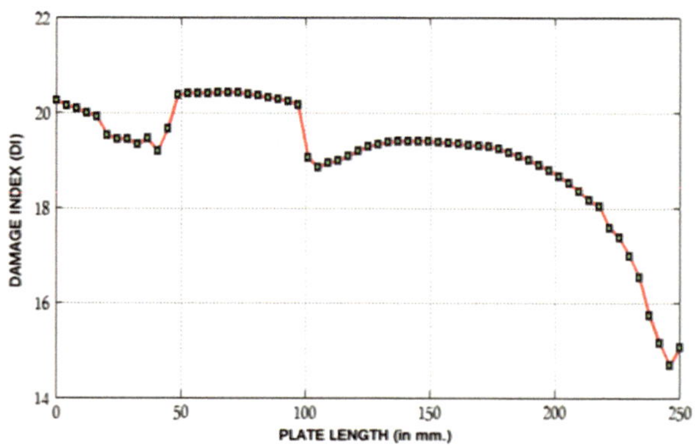

Fig. 8.22 DI of CASE A-2 plate

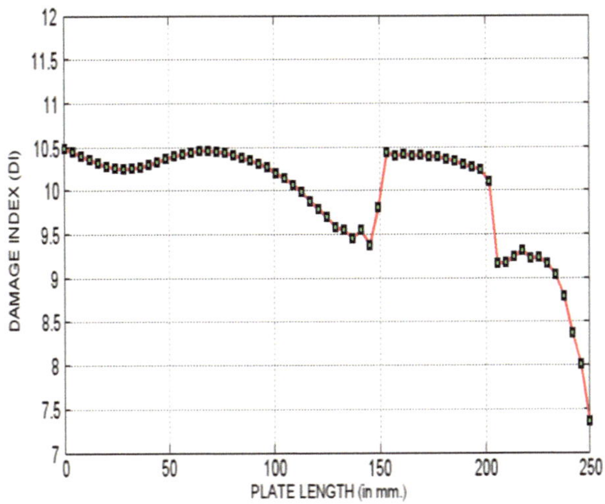

Fig. 8.23 DI of CASE A-3 plate

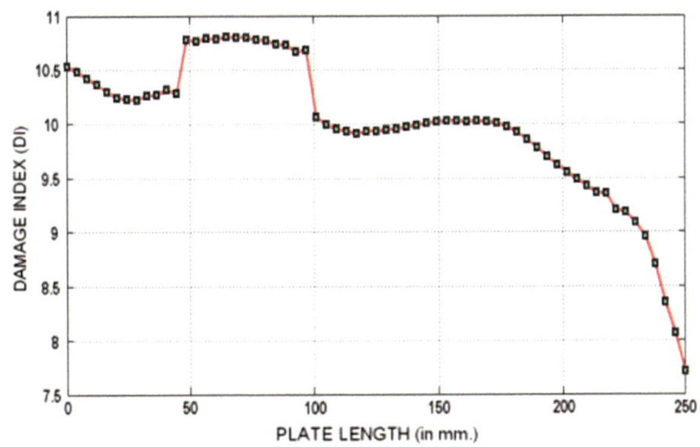

Fig. 8.24 DI of CASE B-1 plate

8.9 Summary

In this chapter, we have first discussed the concept of structural health monitoring and its application in various fields of engineering. We have then considered the application of SHM for a generic case of Actuation and Sensing in Composite Laminate. We have developed the related theory toward this direction at length. Subsequently, we have talked about the development of SHM algorithm based on feature collection and sensor data fusion. Finally, we have shown that for a typical damage like delamination in composite what features can be considered and the relative performance of each of the features in predicting damage is discussed with the help of a numerical example.

Model Questions:

1. Describe the significance of SHM in various engineering fields.
2. What are the basic axioms of SHM? Draw a flow diagram showing interrelationship between the axioms.
3. What factors are important for the selection sensors and why?
4. Find out how SHM can be applied for vibration control of a symmetric cantilever laminated composite beam.
5. What is multi-sensor data fusion? Discuss the advantage of Omnibus technique of data fusion over Waterfall technique.
6. What are the most important features for damage detection in laminated composite plates? Write short notes on each of the features.

References

1. Worden K, Farrar CR, Manson G, Park G (2007) The fundamental axioms of structural health monitoring. Philos Trans Royal Soc A 463:1639–1664
2. Balageas D, Fritzen C-P, Güemes A (2001) Structural health monitoring. ISBN: 978-1-905209-01-9
3. Farrer CR, Lieven NAJ (2007) Damage prognosis: the future of structural health monitoring. Philos Trans Royal Soc A 365:623–632
4. Worden K, Dulieu-Barton JM (2004) An overview of intelligent fault detection in systems and structures. SHM 3(1):85–98
5. Liang W, Chan THT (2009) Review of vibration-based damage detection and condition assessment of bridge structures using structural health monitoring. In: The second infrastructure theme postgraduate conference: rethinking sustainable development: planning, engineering, design and managing urban infrastructure, March 26, Queensland University
6. Jaiswal AK, Kumar A, Bhattacharya B (2008) Structural health monitoring of ribbon reinforced composite laminate using piezoelectric sensory layer. Int J COMADEM 11:09–17
7. Adams RD, Walton D, Flitcroft JE, Short D (1975) Vibration testing as a nondestructive test tool for composite materials. In: Composite reliability ASTM STP 580, American Society for Testing and Materials, pp. 159–175
8. Cawley P, Adams RD (1979) The location of defects in structures from measurements of natural frequencies. J Strain Anal 14(2):49–57
9. Salawu OS (1997) Detection of structural damage through changes in frequency: a review engineering structures 19(9):718–723
10. Wahab MMA, Roeck GD (1999) Damage detection in bridges using modal curvatures: application to a real damage scenario. J Sound Vib 226(2):217–235
11. Li YY, Cheng L, Yam LH (2002) Identification of damage locations for plate-like structures using damage sensitive indices: strain modal approach. Comput Struct 80:1881–1894

Chapter 9
Future Directions of Pipe Investigating Robots (PIRs)

9.1 Introduction

In this chapter, we will provide some of the future directions in the field of Pipe Investigating Robots. Starting from very simple cable-drawn vehicles, extensive research has enabled the transformation of the passive system into an autonomous wireless monitoring system. However, there are many challenges and scope of multidisciplinary research in this field. This includes new research efforts from mechanical engineering, electronics and communication system design, computer science in terms of data mining and software development, and materials science for new sensor development. We will begin with the issue of advanced robot development for very small conduits. Subsequently, we will discuss the scope of improvement in wireless communication system. Next, we will discuss the design of new sensors and finally, we will touch the commercial aspect of developing such robots specifically in the context of software development for better user interface.

9.2 Future Directions in Advanced Robot Development

Currently, pipe inspection robots are used for gas and water pipeline inspection works. The diameters of these pipes are reasonably large, and hence, wheeled robots are suitable for such applications. However, in future, the requirements of pipe-crawling robots may expand to the field of monitoring of biological systems such as endoscopy. The relevant PIR getting developed toward this direction is in the form of micro-crawler robots. Considering several bio-inspired designs in this direction, the two of the most potentially interesting mechanisms are based on bio-inspired **Peristalsis Motion** and **Inchworm Omega-shaped motions,** respectively.

H. Ogai and B. Bhattacharya, *Pipe Inspection Robots for Structural Health and Condition Monitoring*, Intelligent Systems, Control and Automation: Science and Engineering 89, https://doi.org/10.1007/978-81-322-3751-8_9

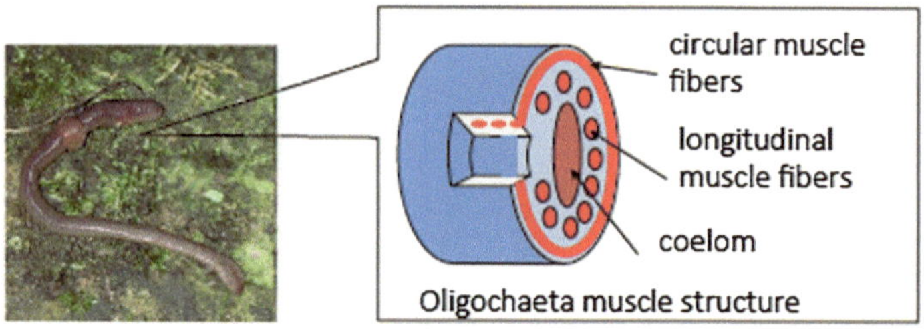

Fig. 9.1 The muscle structure of Oligochaeta

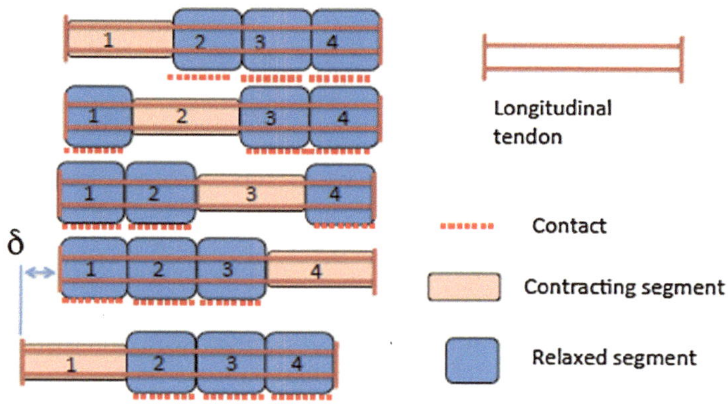

Fig. 9.2 The gait sequence of the worm

Worms of species Oligochaeta are known to deform their body for generating forward motion gait. Figure 9.1 shows the muscles employed by this species to generate the forward motion [1].

With a sequence of contraction of radial muscles, each individual parts of the body gets detached from the ground and a traveling wave is generated on the worm body. As explained in Fig. 9.2, this wave motion finally culminates in the longitudinal expansion of the first unit and creates positive displacement.

Fig. 9.3 The Giant Looper moth in motion

This type of motion generation requires the deployment of two antagonistic smart actuators. For example, shape memory alloy-based springs are used to develop such motion.

Another bio-inspired process is the development of Omega-type inchworm motion shown by Ascotis selenaria, commonly known as the 'Giant Looper.' Figure 9.3 shows the caterpillar of Giant Looper moth, which forms the Omega-like structure during forward motion.

This type of motion generation requires multiple links with smart composite flexure and SMA coil for contracting and relaxing periodically. The front leg is specially developed such that it has sticky pads on one side. This hinders backward movement of the system. The backside leg, on the other hand, is free to move. The complete mechanism is shown in Fig. 9.4.

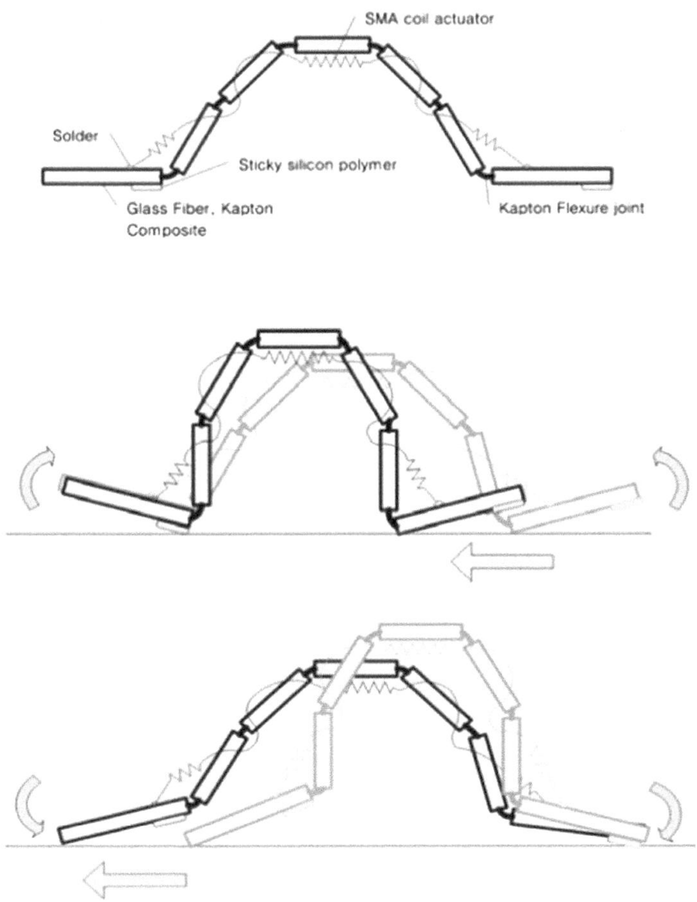

Fig. 9.4 Inchworm motion using SMA actuation [2]

9.3 New Wireless Communication System Design

There are many commercially used pipe inspection robots. But almost all of these robots are wired systems. The reasons are generally as follows:

- The range of the radio signal transmission in pipe is short and weak in strength.
- The robot needs source of stored energy like battery power for motion. If the battery of the robot gets over, the robot may stop anywhere inside the pipe.

Hence, the development of wireless communication system is a challenging one for inside pipe inspection robot. This may initiate research in the following directions:

(a) Long distance steel pipe wireless communication system

For the long distance steel pipeline, the combination of wireless and wired communication systems is shown in Fig. 9.5. The inspection robot in the pipe communicates with the nearest access point (AP) on the pipe and the AP is connected with a router on the ground. From this router, the PC of the base station is connected by wireless communication system. Thus, there is a combination of wireless–wire–wireless communication system employed for long distance coverage.

(b) Long distance ceramic pipe wireless communication system by using booster

Using wireless LAN and general purpose device to reduce radio wave absorption in sewers is a possible way to enhance the range of communication distance. The following actions are necessary for this purpose.

1. Use a frequency band with less attenuation. Thus, 5 GHz may be a better solution in comparison to 2.4 GHz band for 30 cm diameter pipe network.
2. There should be a cleansing robot for washing the tube wall to remove the sludge and dry it. This increases the communication efficiency significantly.
3. Use of coaxial cable with low attenuation and antenna with large gain are other remedies for this problem (Fig. 9.6).

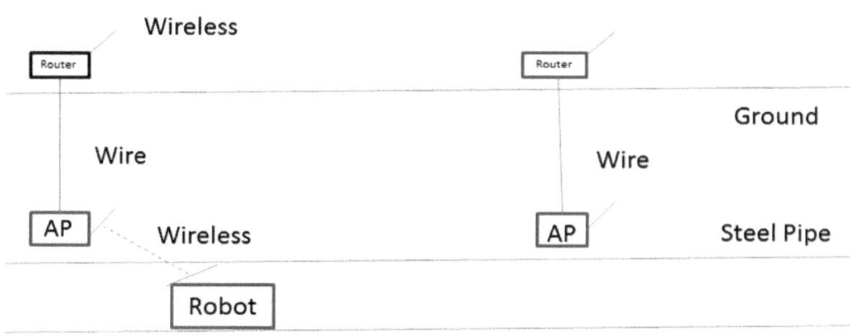

Fig. 9.5 Wireless communication system for steel pipe inspection robot

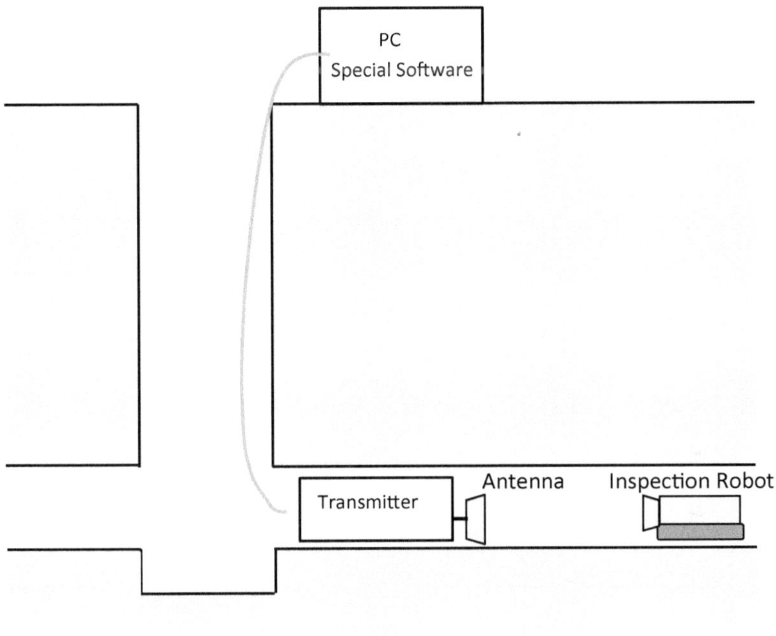

Fig. 9.6 Advanced radio frequency transmission system

4. By using a dedicated radio transceiver, the transmission power may be increased and the communication range can be enhanced.

A booster (amplification repeater) as shown in Fig. 9.7 fixed over a booster robot (Fig. 9.8) can be employed for this purpose.

(c) Wireless power transmission by microwave

For the battery loss in the pipeline, the wireless power transmission technology could be useful in future.

Fig. 9.7 Booster of RF signal

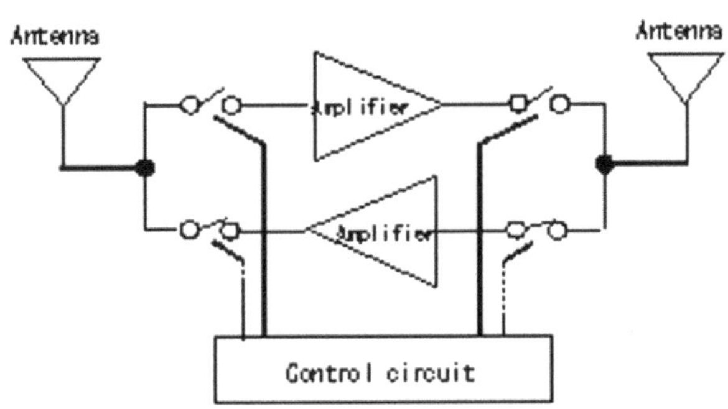

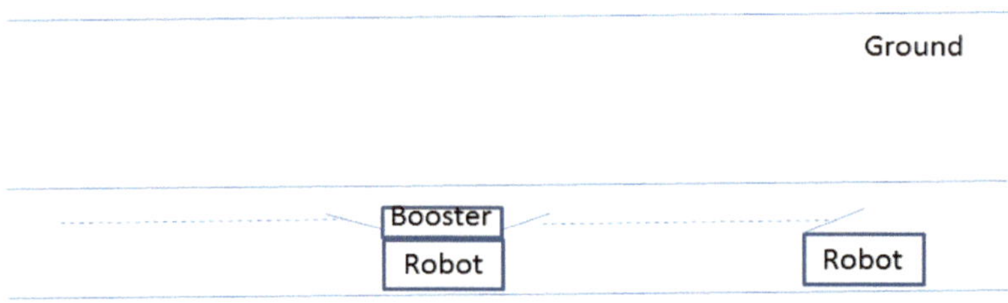

Fig. 9.8 Long distance ceramic pipe wireless communication system with mobile booster

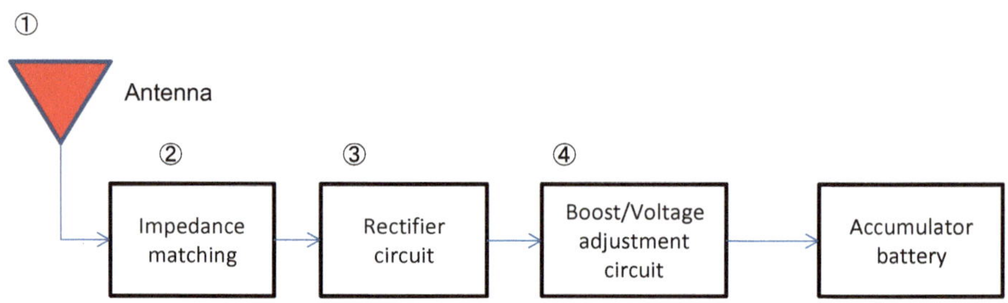

Fig. 9.9 Wireless power transmission system (power receiver)

A small experiment was carried out for some power transmission of wireless sensor network system by microwave system at the IPS Waseda University. The system is clearly explained in Fig. 9.9.

①, ②, ③ is a device in which the antenna and the rectifier circuit are unitized and is called a rectenna. Regarding ④, it is a circuit capable of boosting input power of several tens of μW.

As a result, a power transmission distance of about 10–20 m can be realized.

In future, wireless data and power parallel transmission technology are expected to be developed simultaneously for more intelligent pipe inspection systems.

9.4 Advanced Sensors

Many advanced sensors are getting developed today for more robust and accurate monitoring of pipe health. We will discuss two such sensors, PVDF probe and LED–LDR scanner here.

9.4.1 PVDF Probe Sensor

As discussed in Chap. 7, PVDF sensor works on the principle of piezoelectricity, i.e., when the stress is applied to the flexible probe, it generates electrical charges.

This generation of electric charge is directly proportional to the load/stress applied. Figure 9.10 shows a probe based on PVDF. Experiments are conducted using this probe for detecting cracks or any non-uniformity on the sample surface. This probe consists of two major parts, i.e., the tactile part and the piezoelectric beam. Tactile part of the sensor will always be in contact with the surface of the pipe. The piezoelectric beam consists of three layers. The first layer is made up of metal and acts as a host cantilever beam, the second layer is an electric insulator, and the third layer is a PVDF strip. PVDF strip has a significant operational bandwidth of 0–20 GPa for pressure (force) sensing, the capacitance of 750 pF, dielectric constant of 2000, electromechanical coupling factor of 60, piezo-constant of 6–7 pC/N and impedance of 1 MΩ with operating frequency close to 100 Hz. Whenever the tactile part experiences irregularity in the surface over which the probe is moving, it deflects the beam, which in turn exerts pressure on PVDF. With increasing deflection, the dynamic stress will increase and hence more electric charges will be generated in the transducer. These electric charges are converted into voltage signals and are further utilized for damage detection.

Experiments are conducted for two different types of motion of the rotating probe, i.e., uni-directional and bi-directional analysis. In the uni-directional analysis, the probe is rotated in one direction only, and the resulting data are analyzed. However, in the bi-directional motion, the probe is rotated in both forward and backward direction. It is observed that bi-directional analysis shows better results as compared to the uni-directional analysis. This is because in the case of uni-directional analysis, the probe always deflects in a particular direction leading to permanent deformation of the probe; however, in bi-directional analysis, direction of deformation is in opposite directions during clockwise and anti-clockwise rotation, respectively.

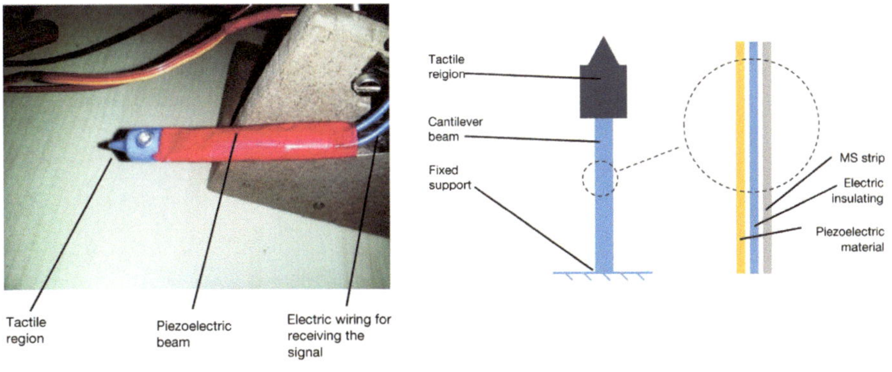

Fig. 9.10 A piezoelectric PVDF probe for pipe health monitoring

9.4.2 Opto-electric Sensors

Light is used as the sensing tool in this technique. Light from the LED travels through the medium (fluid) and gets reflected back from the pipe surface. The reflected light falls on the LDRs. The lights sensitive LDR will respond according to the quantity of light that falls on it. Cracks in the pipeline will make less amount of light to fall on the LDR making it less resistive. The signal acquired from the LDR will have changes embedded in it based on the amount of light incident on it. The optical sensor consists of a transmitter and receiver. Transmitters are the Light-Emitting Diodes (LEDs) which constantly emit light, and the receivers are the Light Dependent Resistors (LDRs). The LEDs convert electrical energy into optical energy (light energy) [3, 4]. LDR is a device whose resistivity is a function of the incident electromagnetic wave. LDR device works on the principle of photoconductivity. When light falls, i.e., photon falls on the device, the resistance of the LDR changes. The resistance of the LDR decreases with increase in incident light intensity and increases in less light. In what follows is the description of an experiment carried out in this direction.

The LEDs and LDRs are mounted on a turbine whose diameter is 170 mm. There are totally twenty LDRs mounted on the circumference of the turbine, and distance between the two LDRs is 24 mm as shown in Fig. 9.11. The LEDs are powered by a 12 V battery supply. The LDRs are supplied 5 V through a microcontroller. The general parameters shown in Table 9.1 were used for acquiring sensor signals.

A microcontroller acquires the signal from LDRs and feeds it to LabVIEW software in an adjunct computer. LabVIEW is used for the data acquisitions and the data storage in the PC. An Experimental set-up for calibrating the LED-LDR based sensing is shown in the Fig. 9.12.

A lead screw is used for converting rotary input motion into linear output motion. The turbine is connected with the lead screw with the help of a lead nut.

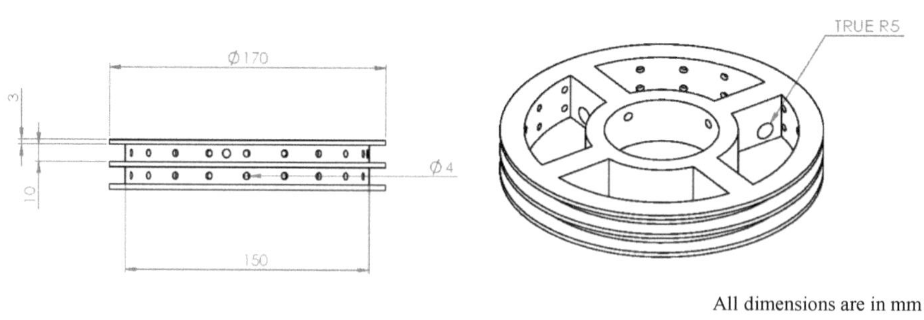

Fig. 9.11 Opto-electric turbine geometrical view

Table 9.1 General parameters

S. No.	Parameters	Values
1	Lead screw length in mm	590
2	Pitch of screw in mm	5
3	Diameter of turbine in mm	170
4	Sampling frequency in Hz	8

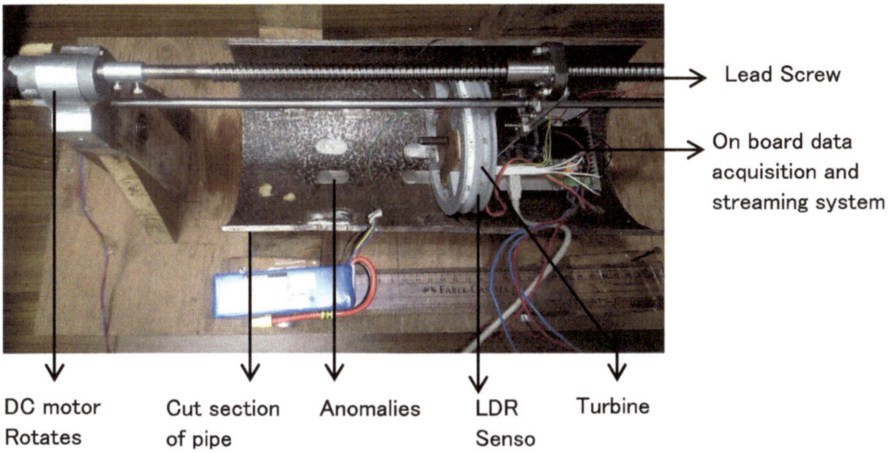

Fig. 9.12 Experimental setup for opto-electric sensors

The nut is constrained from rotating with the screw, so as the screw is rotated the nut travels back and forth along with the turbine along the length of the shaft as shown in Fig. 9.12.

The rotary motion of lead screw is controlled by the DC motor which is connected on one end of the lead screw. LabVIEW software is used to control the angular speed and direction of DC motor. This facilitates the control of the linear velocity of turbine.

Initially, the DC motor rotates at 35 rpm and the corresponding linear velocity of turbine is about 3 mm/s. When the DC motor starts to rotate at a particular rpm, the turbine also moves along the length of the screw maintaining a sensor liftoff of 20 mm. When the turbine starts to move in axial direction of the pipe, the LEDs emit light and LDRs receive the reflected light. The signal generated by the LDR is continuously stored in the PC with help of microcontroller and LabVIEW software. Similarly, the experiment may be conducted for different liftoff distances of LDRs and different linear velocities of the turbine. Figure 9.13 shows a typical LDR output with respect to time. For uniform turbine velocity, this can be easily converted to spatial mapping of the system.

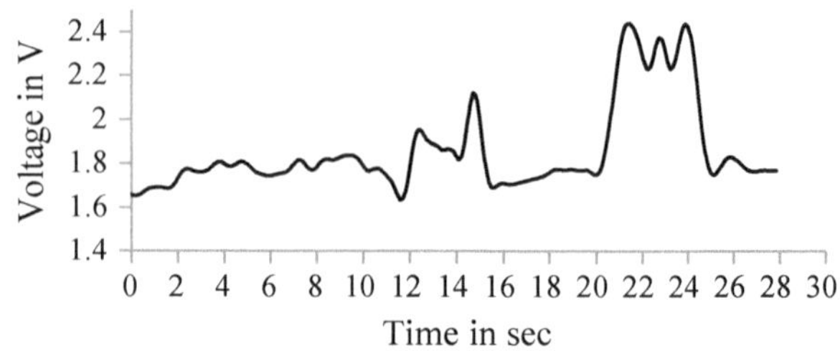

Fig. 9.13 A typical profile generated by LDR. The peaks denote the positions where pipe has cracks

9.5 Software Development

Finally, in order to improve the usability of the system, we need to develop software architecture for suitable user interface. The following are the key architectural layers to be considered for this purpose.

Key architecture layers

- **File Systems**—Distributed file systems which provide storage, fault tolerance, scalability, reliability, and availability.
- **Data Stores**—Evolution of application databases into Polyglot storage with application-specific databases instead of one-size-fits-all. Common ones are Key-Value, Document, Column, and Graph.
- **Resource Managers**—Provide resource management capabilities and support schedulers for high utilization and throughput.
- **Coordination**—Systems that manage state, distributed coordination, consensus, and lock management.
- **Computational Frameworks**—A lot of work is happening at this layer with highly specialized compute frameworks for Streaming, Interactive, Real Time, Batch, and Iterative Graph (BSP) processing.
- **Data Analytics**—Analytical (consumption) tools and libraries, which support exploratory, descriptive, predictive, statistical analysis, and machine learning.
- **Data Integration**—These include not only the organization tools for managing pipelines but also metadata management.
- **Operational Frameworks**—These provide scalable frameworks for monitoring and benchmarking.

Along with pipeline safety data, information safety is also important for any industry. For maintaining privacy and transparency in the system, one needs to create a hierarchy-based information access system using user login module. Account activation can only be done by the admin of software that is pipeline operator by default. Figure 9.14 shows a typical front page of the software developed at IIT Kanpur.

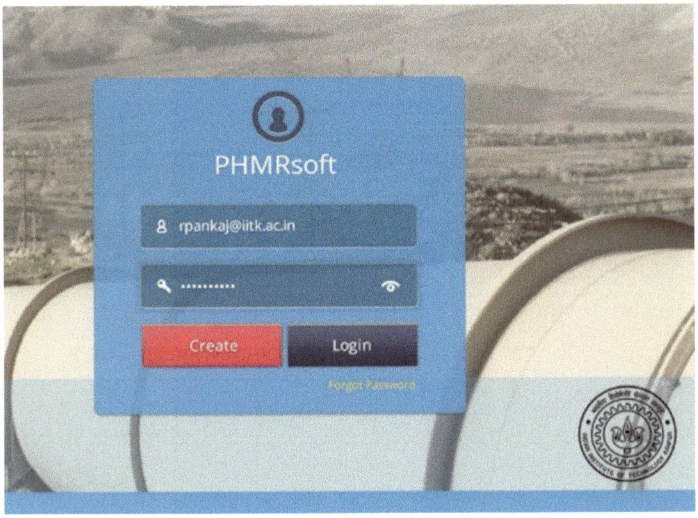

Fig. 9.14 Front page of a pipe health monitoring software

It can easily integrate it with company's intra-industry network.

To support fast access to pipeline data, we create a centralized repository that can store and retrieve the long-range pipeline data with fast speed, and we could integrate inspection data with all related pipeline information to support effective decision making and ensure pipelines are operated safely and cost-effectively (see Fig. 9.15).

Fig. 9.15 Typical pipeline data storage system architecture

Fig. 9.16 A typical pipeline network visualization system

The user can also review previous inspection reports, and it supports a variety of formats and scale with data alignment and segmentation facility. Figure 9.16 shows a typical layout of the system.

For inspection data analysis, it can provide all necessary information related to the inspected pipeline as design parameters, operational parameter, historical data, and risk-based prediction of potential threats. It also provides routes and schematic information of pipelines. The software can keep a record of the complete inspection

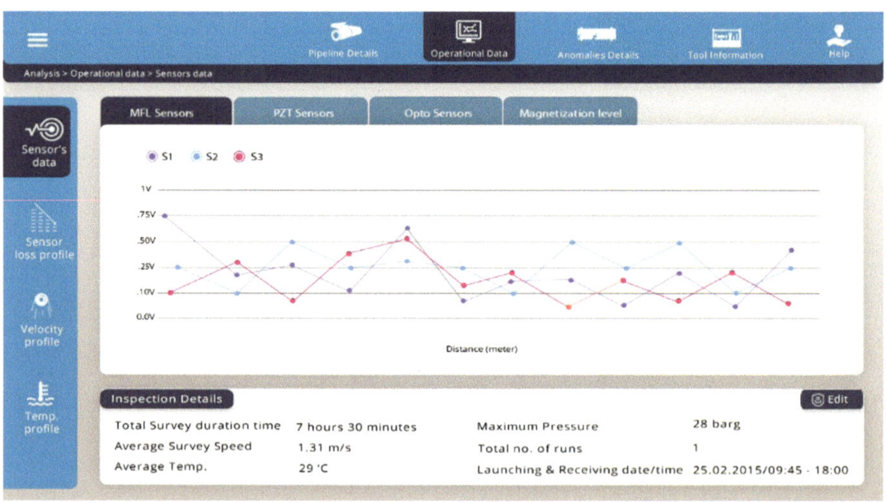

Fig. 9.17 A typical inspection result layout

process in the form of snaps and videos which could be used for training and other application.

Finally, to evaluate inspection data, we need information regarding sensors and tools. In the proposed layout in Fig. 9.17, one can access tools information along with sensor data in an interactive manner. With the help of sensor loss profile, temperature, velocity profile along with operator's criteria, we can quickly evaluate whether inspection results are satisfactory or not.

9.6 Multi-Axis Robots

As shown in Fig. 9.18, the mechanical structure of every pipeline robot has followed in the same mechanical design principle, at the same time, pipeline robot has different functions in every generation.

The pipeline robot ver. 1 owns the most basic function; it can move in pipeline according to the signal which is sent by human through cable; the fixed camera can record the video information of pipeline's inside situation, at the same time, it is also conformal to the basic mechanical structure of pipeline robots developed at the Ogai laboratory. As the figure shows, for all the following robots, mechanical structure is designed based on the original structure. It has added new structure and function on robot ver. 2, in order to fit pipeline which has different diameters [5–10]. There has been the new stretching structure which is used to change the usage range of the robot. There has been also been added pressure sensors in the four wheels so that it can receive the information of pressure sensed by the four wheels. Based on robot ver. 2, there has been added an automatic running system in robot ver. 3. By infrared sensor which is fixed on the robot, it can move in pipeline without human intervention; thus these robots have become more intelligent.

(a) The mechanical structure of newest pipeline robot

As shown in Fig. 9.19, extending the former pipeline robots, there has been a new mechanical structure for robot ver. 4.

Fig. 9.18 Development of Ogai laboratory's pipeline robots

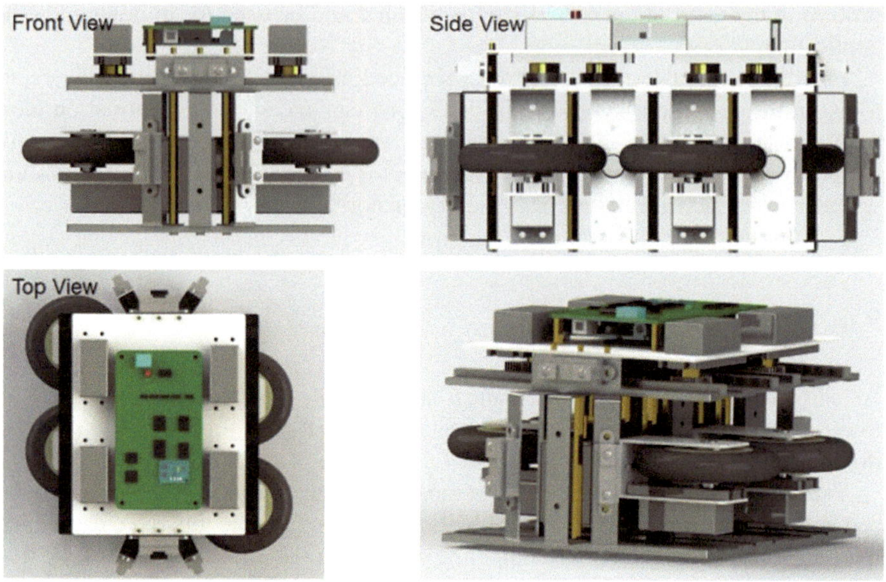

Fig. 9.19 The new mechanical model of robot ver. 4

Because of the size of servo motor, in the former pipeline robots, one needs enough inside space to fix the servo motors. The four servo motors, which are fixed inside the robot, take a large amount of inside space; consequently, all former pipeline robots are bigger in size. Thus, the robot is useless under the situation of monitoring pipelines with smaller diameter. In order to avoid the obvious disadvantage, there has been a new mechanical structure in robot ver. 4. As shown in Fig. 9.20, the fixture posture of servo motor has been changed, at the same time, there have been the new arrangement mode for four wheels, different from the

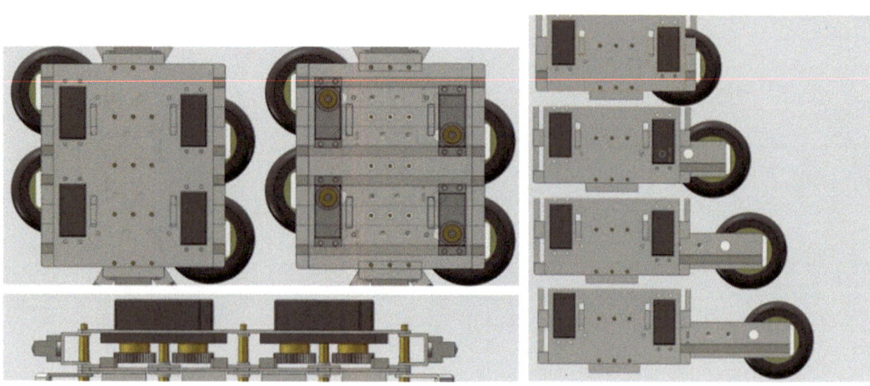

Fig. 9.20 The stretching system of new pipeline robot

symmetry mode, four wheels has been arranged in cross mode. By using the new arrangement, the inside space of the robot has become more compact and the robot ver. 4 has become smaller in size.

As the former version, there is also a stretching system in robot ver. 4, as shown in Fig. 9.20. The main function of stretching system is to control the action of extension and retraction of four driven wheels. In this system, it needs to change the rotation of servo motor into the lateral movement of driven wheels; hence, we use the gear-rack system to achieve the motion conversion.

Because of the cross arrangement mode, there has been enough inside space for the stretching system. The robot ver. 4 with bigger usage range (diameters 200–400 mm) can fit more diameters of pipeline than the former robot version with smaller usage range (diameter 250–300 mm).

(b) The function of the latest pipeline robot

Similar to the robot ver. 3, the ver. 4 also has an automatic motion system. The automatic motion system is achieved based on the distance information from the infrared sensors. Also, by using the wireless control module, the robot can be controlled by human in manual mode. In the sensor front, there have been added wireless video modules, which can produce real-time video information when we use the wireless control mode. The ver. 4 robot can scan in the running mode with higher precision; as a new module MPU6050 is added in robot ver. 4. By the use of this module, the posture change of robot can be recorded while running in the pipeline. According to the posture information of the robot, it can predict the trend of pipeline. It is meaningful for some pipelines for which the network design is not available.

(c) Automatic running system

As shown in Fig. 9.21, the automatic running system of the robot consists of different parts.The Arduino Mega 2560 is the core control part or the brain of this robot. The main role of it is to make comparison and send instructions to servo motor. The main role of infrared sensor is to collect the distance information. It is the most essential part in the automatic running system. The battery can supply power to each and every part. In a general situation, the Arduino board can make a

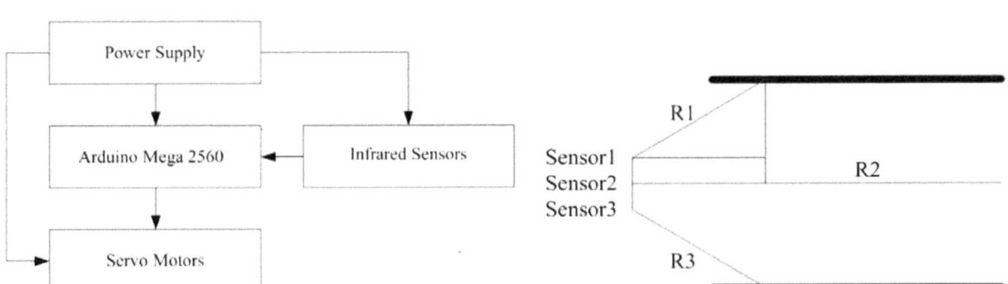

Fig. 9.21 The automatic running system design

comparison with different distance information received from infrared sensors which are fixed in front of the robot: R1, R2, R3 are the distance information from three different sensors. By comparing these different distance information, the core control board can make a judgment about the nature of pipeline in front of the robot. Subsequently, it can send related instruction to the servo motor to control the size of the robot by controlling the stretching system.

9.7 Global Summary

In this chapter, we have explained the various directions and scopes of development of the pipe health monitoring robots. Initially, we have talked about new mechanisms for smaller pipe diameters. The micro-crawler systems are essentially based on smart actuators and smart composite joints. Later, we have talked about the lacuna in the communication design system and ways to improve the communication design for both steel and ceramic pipes. We have then discussed two advanced sensors: PVDF and LED–LDR array which are being developed for better sensing of pipe faults. A layout of the software development is then discussed, and finally, we have discussed the hierarchical development of advanced robots at the IPS Waseda University.

Model Questions

1. Describe one new mechanism that can be used for smaller diameter conduits.
2. How can you expand the wires transmission distance in the steel pipe or ceramic pipe?
3. Describe how to configure the wireless system to achieve both wireless communication and wireless power transmission.
4. Describe a new sensing mechanism for pipe fault detection.
5. Describe how to develop adoptive robot for pipes of different sizes as discussed in the case of multi-axis robot.
6. What in your opinion will be the most innovative development in Pipe health monitoring robot?

References

1. Seok S, Onal CD, Wood R, Rus D, Kim S (2010) Peristaltic locomotion with antagonistic actuators in soft robotics. In: 2010 IEEE international conference on robotics and automation (ICRA). doi:10.1109/ROBOT.2010.5509542
2. Koh JS, Cho KJ (2010) Omegabot: crawling robot inspired by *Ascotis selenaria*. In: IEEE international conference on robotics and automation. Anchorage, Alaska, USA, 3–8 May 2010

3. Held G (2009) Introduction to light emitting diode technology and applications. CRC Press, Boca Raton
4. Mottier P (2009) LED for lighting applications, 3rd edn. Wiley, New York
5. You W, Dongmei Wu, Ogai H, Hirai K, Abe T, Sato G (2008) Wireless radio communication system for a pipe inspection robot. In: The 13th international symposium on artificial life and robotics, pp 393–396, 31 Jan, 2 Feb 2008
6. Ogai H, Yeh Y, Wu D, Hirai K, Abe T, Sato G (2009) Pipe inspection robot using wireless communication system. In: The 14th international symposium on artificial life and robotics, pp 59–62, 5–7 Feb 2009
7. Wu D, Ogai H, Yeh Y, Hirai K, Abe T, Sato G (2009) Pipe inspection robot using a wireless communication system. Airtif Life Robot 14(2):154–159
8. Yang HP (2011) Development of the pipe inspection robot with flexible movements. Thesis of Master of Engineering, Waseda University
9. Kang CC, Ogai H (2015) Development of the pipe robot with flexible movements and adaptive diameter. In: The twentieth international symposium on artificial life and robotics 2015 (AROB 20th 2015), B-Con Plaza, Beppu, Japan, pp 350–355, 21–23 Jan 2015
10. Gu Z, Ogai H, Tang Z (2016) Autonomous navigation pipe robot with infrared sensors. In: Automatic control joint conference, pp 1–6, Kokura, Kitakyushu City